FREEZE–DRYING OF FOODS

Author:
C. JUDSON KING
Department of Chemical Engineering
University of California
Berkeley, California

published by:

A DIVISION OF
THE **CHEMICAL RUBBER** CO.
18901 Cranwood Parkway · Cleveland, Ohio 44128

This book represents information obtained from authentic and highly regarded sources. Reprinted material is quoted with permission, and sources are indicated. A wide variety of references is listed. Every reasonable effort has been made to give reliable data and information, but the author and the publisher cannot assume responsibility for the validity of all materials or for the consequences of their use.

CRC MONOSCIENCE SERIES

The primary objective of the CRC Monoscience Series is to provide reference works, each of which represents an authoritative and comprehensive summary of the "state-of-the-art" of a single well-defined scientific subject.

Among the criteria utilized for the selection of the subject are: (1) timeliness; (2) significant recent work within the area of the subject; and (3) recognized need of the scientific community for a critical synthesis and summary of the "state-of-the-art."

The value and authenticity of the contents are assured by utilizing the following carefully structured procedure to produce the final manuscript:

1. The topic is selected and defined by an editor and advisory board, each of whom is a recognized expert in the discipline.

2. The author, appointed by the editor, is an outstanding authority on the particular topic which is the subject of the publication.

3. The author, utilizing his expertise within the specialized field, selects for critical review the most significant papers of recent publication and provides a synthesis and summary of the "state-of-the-art."

4. The author's manuscript is critically reviewed by a referee who is acknowledged to be equal in expertise in the specialty which is the subject of the work.

5. The editor is charged with the responsibility for final review and approval of the manuscript.

In establishing this new CRC Monoscience Series, CRC has the additional objective of containing the ever-rising cost of publishing, and scientific publishing in particular. By confining the contents of each book to an in-depth treatment of a relatively narrow and well-defined subject and exercising rigorous editorial control, the publishers have ensured that no irrelevant matter is included.

Although well-known as a publisher, CRC now prefers to identify its function in this area as the management and distribution of scientific information, utilizing a variety of formats and media ranging from the conventional printed page to computerized data bases. Within the scope of this framework, the CRC Monoscience Series represents a significant element in the total CRC scientific information service.

B. J. Starkoff, President
THE CHEMICAL RUBBER Co.

This book originally appeared as part of an article in *CRC Critical Reviews in Food Technology*, a quarterly journal published by The Chemical Rubber Co. We would like to acknowledge the editorial assistance received by the Journal's editor, Thomas E. Furia, CIBA-GEIGY Corp. Dr. Theodore P. Labuza, Associate Professor of Food Engineering, Massachusetts Institute of Technology, Cambridge, served as referee for this article.

AUTHOR'S INTRODUCTION

The purpose of this review is twofold—to give a general overview of the status of knowledge and techniques for freeze-drying as used for foods, and to update previous reviews of this field by giving a more intensive coverage of the progress of the last ten years. Emphasis is given to engineering factors and to the fundamental understanding of freeze-drying which must underly sound and intelligent engineering design and process innovation.

I want to express my appreciation for the innumerable and intangible contributions which have come from students in the Department of Chemical Engineering of the University of California at Berkeley who have worked with me in freeze-drying, including Orville Sandall, Robert Gunn, Peter Clark, S. K. Chandrasekaran, Richard Bellows, Thomas Triebes, Wing Lam, Argyrios Margaritis, Romesh Kumar, Peter Cheng, Paul Dhillon, and Gilberto Guzman. I also want to express my thanks to the Western Utilization Research and Development Division of the Agricultural Research Service of the United States Department of Agriculture, who have supported my endeavors in freeze-drying since 1964.

C. Judson King
Berkeley, California

THE AUTHOR

C. Judson King is Professor of Chemical Engineering and Vice-Chairman of the Chemical Engineering Department, University of California, Berkeley.

He received degrees in Chemical Engineering from Yale University in 1956 (B.E.), and Massachusetts Institute of Technology (S.M., 1958, and Sc.D., 1960).

Dr. King has taught at M.I.T. and Berkeley and has consulted for or been employed by Humble Oil and Refining Co., Oak Ridge National Laboratory, Amicon Corp., and Procter and Gamble Co., among others.

He is the author of *Separation Processes*, McGraw Hill, New York, 1971, and of about 45 papers in technical journals.

TABLE OF CONTENTS

INTRODUCTION

Freeze-drying is the most prominent example of separation by sublimation. In freeze-drying water is removed as vapor from a frozen substance. The water passes from the solid phase directly into the vapor phase without becoming a liquid *en route;* consequently, it is necessary that the temperature of the sublimation zone in a material being freeze-dried be held below the triple point temperature of the water or aqueous solution in the material being dried.

As a method of dehydration for food preservation freeze-drying, as a rule, produces the highest quality product obtainable by any drying method. Some of the reasons for the high product quality are still being discovered. One prominent factor, however, is the structural rigidity afforded by the frozen material at the surface where sublimation occurs. This rigidity to a large extent prevents collapse of the solid matrix remaining after drying. The result is a porous, non-shrunken structure in the dried product which facilitates rapid and nearly complete rehydration when water is added to the substance at a later time. Other benefits from freeze-drying lie in the low processing temperatures, the relative absence of liquid water and the rapid transition of any local region of the material being dried from a fully hydrated to a nearly completely dehydrated state. This rapid transition minimizes the extent of various degradative reactions which often occur during drying, such as non-enzymatic browning (Maillard reactions), protein denaturation and enzymatic reactions. The low temperatures involved also help to minimize these reactions and reduce those transport rates which control the loss of volatile flavor and aroma species. The lack of prominent occurrence of the liquid state again helps to minimize degradative reactions and discourages the transport of soluble species from one region to another within the substance being dried. In any food material some non-frozen water will almost unavoidably be present during freeze-drying; however, there is often a rather sharp transition temperature for the still-wet region during drying, below which the product quality improves markedly. This improvement of product quality indicates that sufficient water is frozen to give the beneficial product characteristics of freeze-drying.

Despite its capability of providing a very high quality dehydrated product freeze-drying has been and remains an expensive form of food dehydration. A large amount of research and process development in freeze-drying has been carried out in recent years. Several abortive ventures into freeze-drying were made by various firms without the process and product living up to expectations. Improved understanding of freeze-drying has made possible better and more uniform product quality, and the recent and rapid market success of freeze-dried coffee attests strongly to the fact that freeze-drying has indeed arrived as a reliable process yielding a high quality dehydrated product which is attractive to the consumer.

Freeze-drying has been prominent for some time as a procedure in the laboratory for the preservation of biological specimens. During World War II freeze-drying received considerable attention as a means for the preservation of blood plasma; there was also some study of freeze-drying for food preservation at that time, but no substantial use was made of freeze-drying for food dehydration. The development of freeze-drying through World War II is covered by Flosdorf.[1] In the 1950's research was initiated by the Quartermaster Corps of the U.S. Army to develop freeze-drying for military foods, and by the U.S. Department of Agriculture. Research has spread to a vast number of locations in the United States and in other countries around the world. The large number of organizations working on freeze-drying and the large number of developments stemming from this work are evidenced by a recent collection of patents relating to freeze-drying which were issued during the period of 1960 to 1968.[2] Freeze-drying at present finds its largest application in food dehydration; however, other interesting applications have been suggested and explored, including dehydration of radioactive wastes,[3] preparation of porous catalysts,[4] stabilization of free radicals[4] and sublimation of non-aqueous solvents to allow operation in different media at different temperatures.[4]

Reviews

General descriptions of freeze-drying have been given by Rowe,[5] Rey,[6] van Arsdel,[7] and Cotson and Smith.[8] Two comprehensive reviews of freeze-drying research have been published previously, one in 1957 by Harper and Tappel[9] and the other in 1964 by Burke and Decareau.[10] A list of 638 references on freeze-drying published prior to 1963 has been drawn up by the U.S. Department of Agriculture.[11]

More recently, Peck and Wasan have given a review of drying[12] in which some attention is given to freeze-drying. Luikov[13] described his work on freeze-drying in 1966, and Fulford[14] presented a compilation of Soviet work in the general area of drying. A review containing references to recent European research in freeze-drying has been assembled by Meffert.[15]

The purpose of the present review is twofold. First, an effort is made to concentrate upon a few of the most striking papers published in recent years, to relate them to the rest of the field, and to give an evaluation of their findings and the conclusions that they draw. The papers selected for this purpose are listed in the first bibliography tabulation at the end of this review and are referenced by letters rather than by numerals. One criterion for selection of these papers is that they should represent a number of different research groups and a number of different aspects of freeze-drying. The second purpose of this review is to provide a general updating of the previous reviews of Harper and Tappel[9] and Burke and Decareau.[10]

PHYSICAL MECHANISM OF FREEZE-DRYING

The aim of the process designer and of the processor is to provide an economical drying process which gives reliably uniform and high product quality. An understanding of the basic phenomena and mechanisms involved in freeze-drying is essential for this purpose. In this section attention will be given to a qualitative physical description of the freeze-drying process. The ultimate objective of this description is a realistic basis for analyzing *rates* of freeze-drying. The question of drying rates is all important because of the notably long cycle times or residence times which have been required for freeze-drying.

Sharpness of the Frozen Front

Most analyses of freeze-drying rates have been made with the postulation of a sharp and discrete dividing surface between a region which is fully hydrated and frozen, on the one hand, and a region which is nearly completely dry, on the other hand. The front between the frozen and dry zones retreats inward as freeze-drying proceeds. This situation would result if sublimation occurred from a very thin zone near the surface of the frozen region and if the sublimation removed essentially all of the initial moisture from the remaining solid material immediately after the passage of the frozen front from that region. This picture is clearly an idealization, but has been tacitly assumed by most authors to be a reasonable description of the mechanism of freeze-drying. However, a considerable amount of contention has arisen in recent years regarding the degree of sharpness of the retreating frozen front during freeze-drying.

Evidence for a Diffuse Sublimation Front

Meffert[16] measured temperature profiles during the freeze-drying of rutabaga. From the temperatures and the thermal conductivities inferred from them he concluded that some 30 to 40% of the initial water content was left behind by the retreating frozen front. This remaining water had to be removed by secondary drying, or desorption. Bralsford[17] obtained generally good agreement of experimental freeze-drying rates for beef with the model of a sharp and uniform retreating ice front. Nevertheless, he inferred an effect of liquid diffusion which would serve to broaden the transition zone between the fully frozen region and the dry region at higher temperatures of the frozen zone. Thus the model of a continuously retreating sharp frozen zone front would no longer be valid.

Brajnikov et al.[18] carried out an intriguing experiment in which relatively large slabs of beef (90 x 60 x 12 mm) were freeze-dried while receiving heat from a controlled radiant source. The temperature field during drying was measured by use of a series of seven thermocouples, and the sample was cut into layers at various times after the start of drying so that the average moisture content of each layer could be determined. The temperature fields and moisture contents show qualitative agreement with the concept of a uniformly retreating frozen zone. A typical temperature of the frozen zone was $-20°C$, with the surface being held at $60°C$. From their measurements of the moisture contents of different layers at different times, however, Brajnikov et al. infer that there is a transition zone (or volumetric zone of sublimation) between the fully frozen region and the dry region during freeze-drying of beef muscle. Within this zone 94% of the initial moisture is removed from the beef; it passes from 75% moisture to 15% moisture, wet basis. From their results Brajnikov et al. infer a thickness of 1 to 2 mm for this transition zone. It is important to note, however, that the medium was cut into sample layers which were about 1 mm thick for the purpose of the determination of the moisture content. The average moisture content within each layer was determined gravimetrically. Since the layer thickness rivals the thicknesses reported for the transition zone, it is quite possible that the "zone thicknesses" to a large extent reflect the limit of resolution of the method of determining the moisture profile. As is discussed below, non-uniform retreat of a still-sharp ice front could also cause an apparent transition zone for a large enough layer.

Another interesting experiment performed by Brajnikov et al.[18] involved color photography of cut-open specimens of beef which had been freeze-dried once, then rehydrated with a solution of $CoCl_2$, and then partially freeze-dried once again. Divalent cobalt has the property of changing color from pink to blue depending upon the activity of water in its immediate environment. These experiments apparently also indicated the presence of a transition zone; however, it may be necessary to worry about the effect of the cobalt salt in lowering the freezing point, which could cause some liquid to be present at the sublimation front, even at $-20°C$.

The most detailed picture of behavior at

a sublimation front has been reported by Luikov,[c] who observed photographically the sublimation of ice spheres 80 mm in diameter which were placed in a vacuum chamber and which sublimed at an ice temperature of -53 to $-33\,°C$. Luikov observed that ice crystallites of various shapes and lengths (from tenths of a millimeter to several millimeters) formed at the sublimation surface. These crystallites tended to form near surface irregularities. Their growth was observed visually, and it was found that the crystallites could oscillate and rotate rapidly, at speeds up to 70 rpm existing for 20 to 30 revolutions. These oscillations reflect molecular pressure from the escaping vapor. The oscillations could stop and then begin again in the same or the reverse direction, but more often the crystallite would break from the ice surface and be entrained away at velocities on the order of 0.07 to 0.36 meters/sec with the escaping vapor. Luikov[c] and others have pointed out that the very great and very sudden expansion of water from the solid to the highly expanded vapor state upon vacuum sublimation can cause strong forces and unusual flow patterns. Local unevenness of temperature along the sublimation front can also cause mechanical stresses which can cause crystallites to break off and be entrained. Following observations made earlier[19,20] of the entrainment of liquid droplets from the evaporating surface during vacuum evaporation of liquids, Luikov suggests that this entrainment mechanism can have an important effect on the observed rate of sublimation of ice. This would amount to the removal of moisture without the transfer of an equivalent amount of heat to the sublimation front being required.

Luikov[c] also found that sublimation and ablimation (desublimation from the vapor state to the solid state) occur simultaneously at the sublimation front of an ice sphere. This phenomenon again reflects (and strengthens) irregularities in the solid surface which give rise to locally uneven temperatures and water vapor pressures. The overall rate of sublimation must be greater than the overall rate of ablimation, and the net rate of sublimation would be the difference between the two. The occurrence of ablimation was confirmed by

Luikov[c] in experiments involving vacuum sublimation of water vapor from spherical ice crystals made from water containing red fuchsin dye. Crystallites created directly from the solid phase appeared red, but those formed by ablimation after an original sublimation were transparent and colorless. The tendency for ablimation to occur depended upon the degree of saturation of the surrounding gas phase with water vapor. A very low water vapor pressure in the gas discourages ablimation, but ablimation occurs to a greater extent as the surrounding gas becomes more nearly saturated in water vapor. Ablimation can also occur in a supersaturated zone somewhat removed from the ice surface in the vapor phase.

Luikov[c] reports experiments on the drying of beds of quartz sand as a test of the sharpness of the retreating sublimation front during freeze-drying. Local moisture contents were obtained by an unstated method, along with profiles of temperature and total pressure within the sand bed. Experiments were carried out for beds of different porosities, ranging from 0.2 to 0.51. Moisture profiles are presented only for the lowest bed porosity (0.2) and show a gradual change in moisture content from 0 to as much as 15% moisture (wet basis) over a thickness of 30 mm within the sand bed. This is taken as evidence of evaporation occurring at all points within this zone, and thus as evidence of a very large transition zone. Moisture profiles are not reported at the higher porosities, but the pressure and temperature profiles for the higher porosity beds are much more characteristic of what one would expect for sublimation with a sharp ice front. No temperatures or pressures are reported for the case of a bed porosity of 0.2. With decreasing bed porosity the vapor permeability or diffusivity within the bed should decrease while the bed thermal conductivity should increase. By the URIF model (see below) or any other simultaneous mass and heat transfer analysis this means that at decreased bed porosities there will be a tendency for a higher temperature of the frozen zone, with the result that it will be difficult to keep the frozen zone from melting even at moderate outer surface temperatures of the material being dried. The interesting and revealing studies made by Luikov of small-scale behavior on the sublimation front can account for a transition zone with a thickness of perhaps a millimeter, but not for a zone with a thickness of an inch or more. Thus it appears quite possible that the broad transition moisture-content zone reported by Luikov for a bed porosity of 0.2 reflects a large amount of melting during drying because temperatures were not maintained low enough.

Evidence for a Sharp Sublimation Front

Against these indications of a diffuse sublimation front may be marshalled the evidence obtained by various investigators for a relatively sharp sublimation front. A number of writers have reported cutting open partially freeze-dried specimens and observing visually a distinct and relatively sharp demarkation between a still-frozen zone and a dry zone of uniform coloration. These include Hardin,[21] who studied the freeze-drying of beef, Margaritis and King,[22] who studied the freeze-drying of turkey meat, and Beke,[23] who studied the freeze-drying of pork. Beke[23] provides photographs of cut, partially freeze-dried pork samples 14 mm thick after 200, 300, 400, 500 and 600 minutes of freeze-drying. These photographs clearly show a distinct and very sharp transition from a dry zone of uniform appearance to a dark frozen zone, with the frozen zone retreating inward as the drying time increases.

Photographs of partially freeze-dried specimens of white turkey breast meat were obtained by Clark[24] and are shown in Figure 1. The specimens shown were freeze-dried for various lengths of time in the molecular sieve, mixed-bed process,[25,26] described later. The pieces were approximately cubic, 1 cm on a side. After being removed from the freeze-drying process the pieces were subjected to high temperature vacuum oven drying for determination of the remaining moisture content. The samples were then cut in half by means of a razor blade and photographed. A fully freeze-dried sample (Figure 1-g) containing 1.2% of the initial moisture present (2.4%, wet basis) exhibited very little distortion or shrinkage. The other specimens do exhibit shrinkage and distortion which may be attributed to the high temperature vacuum oven drying procedure. The dark, discolored central portion of each of the other specimens represents the region that still contained a high moisture content when freeze-drying was interrupted. The demarkation of this zone of residual moisture is fairly sharp, and the positions of the zones are in agreement with the concept of a continuously retreating sublimation front during freeze-drying. Since melting occurred before the high temperature vacuum oven drying was completed, it should be expected that the residual moisture front shown in the photographs is more diffuse than was actually the case when freeze-drying was stopped, because of the opportunity for liquid migration after the moisture was melted. The demarkations between the frozen and dry zones are sharper in the photographs shown by Beke.[23] In principle, it is possible to compare the volume of the discolored region of each specimen with the percentage of the initial moisture found to be remaining by the vacuum oven analysis. This procedure would be complicated, however, by distortion and shrinkage of the samples through vacuum drying, and the surest conclusions are only that there is a distinct demarkation between a moist core and a relatively dry outer layer, and that there is a moist core remaining even at a residual moisture content of 18% of the initial moisture present (34% moisture, wet basis).

Additional evidence relating to the sharpness of the retreating sublimation front in the freeze-drying of beef was obtained by Hatcher,[27] who used gamma ray attenuation measurements to monitor the profile of moisture content across relatively thick (2-inch) slabs of beef as a function of time during freeze-drying. Radiant heat was supplied to the outer surfaces of the slabs. Hatcher reported that visual inspection of partially dried specimens revealed the ice front to be planar. Since his gamma ray beam diameter was $3/16$ inch, he concluded that the ice front was sharper than could be detected by the degree of resolution afforded by the beam. He also found no indication of residual moisture in an amount sufficient to affect the gamma ray count rate in the dry zone once the ice front has passed

FIGURE 1

Photographs of Partially Freeze-Dried Cubes of Turkey Meat, after Removal of Residual Moisture through High-Temperature Vacuum Oven Drying.[24]

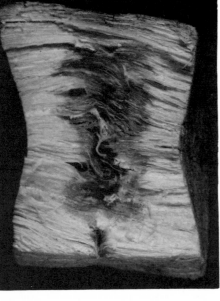

c. Sample 7-1, 27% of initial moisture remaining.

d. Sample 9-4, 21.5% of initial moisture remaining.

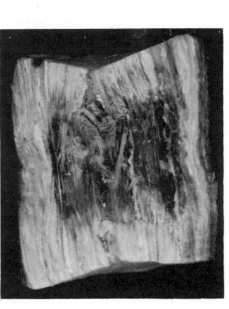

a. Sample 6-1, 36% of initial moisture remaining.

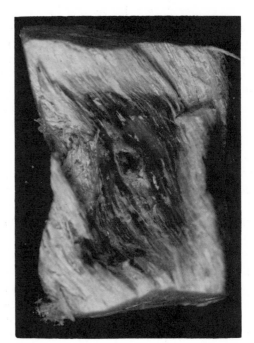

b. Sample 8-1, 33% of initial moisture remaining.

FIGURE 1

e. Sample 5-4, 18.5% of initial moisture remaining.

g. Sample 13-4, 1.2% of initial moisture remaining.

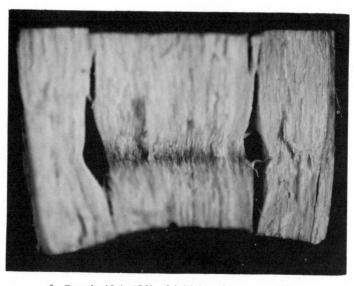

f. Sample 10-1, 18% of initial moisture remaining.

during drying. It does appear from the gamma ray count profiles reported by Hatcher that there may have been some broadening of the transition zone between the dry layer and the frozen layer toward the latter stages of drying; however, this could also be a reflection of nonuniform retreat of a still-sharp ice front, since the gamma ray beam passed through three inches of beef in a direction presumably parallel to the retreating sublimation front. Nonuniform retreat of a sharp ice front has been found by Margaritis and King[22] (see below), and Hatcher does report some preferential drying from the edges of his samples.

Hatcher also reports temperature histories during freeze-drying of the same beef slabs, measured by No. 30 copper-constantan ther-

mocouples threaded into the samples. Figure 2 shows temperatures reported as a function of time at different positions with a 2-inch-thick, 3-inch-diameter beef sample. Figure 3 shows the same temperatures as functions of distance from the outer sample surface at different times after the start of drying. These data indicate a very thin transition between a frozen zone of uniform low temperature, on the one hand, and a dry zone with a nearly linear gradient of temperature between the ice front and the heated outer surface, on the other hand. Results similar to Figures 2 and 3 have been reported by Spiess[65,87] for the freeze-drying of potato starch and egg white, by Sandall et al.[E] for the freeze-drying of turkey meat, by Kessler[46] for freeze-drying of beds of glass

FIGURE 2

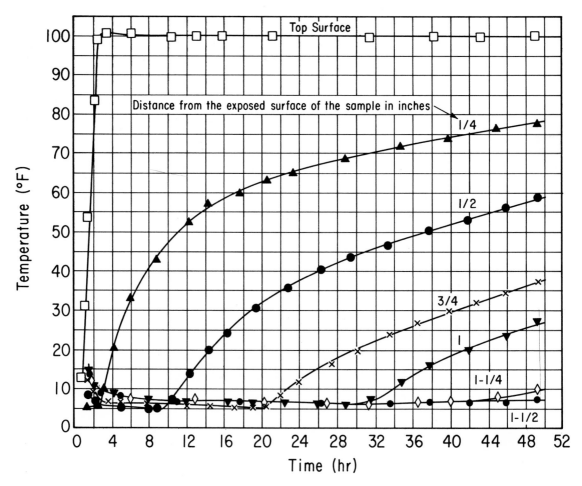

Temperature vs. time at different thermocouple locations during the freeze-drying of beef, after Hatcher.[27] Chamber pressure = 1 torr.

FIGURE 3

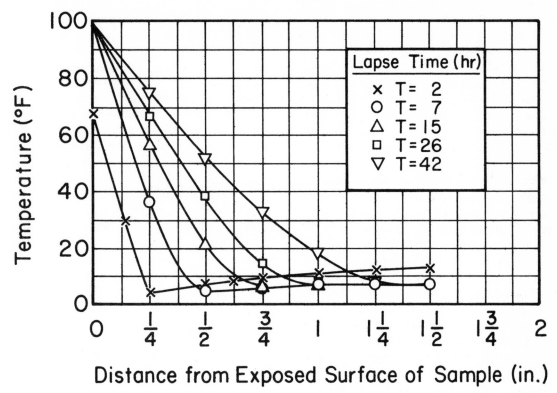

Temperature vs. position at different times during the freeze-drying of beef, after Hatcher.[27] Chamber pressure = 1 torr.

beads and of milk, and by other investigators.

In another interesting experimental approach, MacKenzie[28,29] constructed a "freeze-drying microscope" in which he could observe freeze-drying as it occurred in various transparent frozen liquid systems. The liquid material was cast in a thin layer between flat cover plates and then frozen in place. A vacuum was next imposed upon the specimen from one end so as to cause freeze-drying to occur across the plane of microscope observation. An extremely sharp sublimation front was observed, although the front, while sharp, did not necessarily remain planar.

There must be a finite partial pressure of water vapor present in the gas within the "dry" layer, since the water vapor generated by sublimation must escape across the dry layer. One would therefore expect a finite amount of sorbed water to be present in the dry layer, corresponding at least to the amount that would be predicted by the equilibrium sorption isotherm. Sandall[30] made calculations for typical conditions of freeze-drying of turkey breast meat using sorption isotherms for the same material measured by King et al.[31] For heating from the outer, dry surface he concluded that equilibrium between the dry layer and the escaping water vapor would give an average moisture content within the dry layer equal to 3% of the initial water remaining, or about 1.5% moisture by weight. This surprisingly small value results from the increasing temperatures across the dry layer toward the outer surface, as well as from the decreasing water vapor partial pressure. In a case where heat is supplied across a frozen layer, the average moisture content of the dry layer should be greater since there is not a substantial temperature increase across the dry layer.

Even though the moisture content of the dry layer corresponding to local equilibrium with the escaping water vapor is so low, it would be possible for the average moisture content of the dry layer to be higher if the desorption of bound water from the material being freeze-dried were a rate-limiting step. Margaritis and King[22] assessed the rate of desorption of bound water from the fibers within freeze-dried turkey meat by means of an experiment in which humidified nitrogen was blown through a slab of freeze-dried meat. At the start of an experiment the specimen was allowed to come to equilibrium moisture content with a nitrogen stream of a particular humidity. The humidity of the permeating nitrogen was then changed in a stepwise fashion, and the transient approach of the specimen toward moisture equilibration with the new relative humidity was followed. A capacitance differential pressure gauge was used to monitor the pressure drop across the specimen as a function of time. This instrument had sufficient sensitivity to register the change in pressure drop caused by swelling and shrinkage of the fibers of the meat as moisture was sorbed or desorbed. The gas flow was rapid enough so that the equilibration process was not controlled by moisture supply in the gas and slow enough so as to avoid distortion of the mounted sample. Desorption equilibration was found to be largely complete in times less than an hour. Since freeze-drying generally takes a number of hours to be completed, it was concluded that the desorption of bound moisture from meat fibers is not a likely rate-limiting step, and hence the dry layer in a freeze-drying process should be expected to be close to local equilibrium with the vapor phase. As a result the moisture content of the dry layer, once the sublimation front has passed, should be quite low in a process in which heat is supplied from the outer, dried surface.

Yet another justification for postulating a sharp sublimation front is the agreement of the thermal conductivities inferred by Sandall et al.[E] and Bralsford[17] by application of the uniformly retreating ice front model to their drying data with thermal conductivities measured for the same specimens independently when dry (see below). It is known[32] that the thermal conductivity of freeze-dried meats increases as the moisture content increases, so that an appreciable moisture content left after the passage of the sublimation front should cause higher thermal conductivities to be inferred from drying data than would be measured for totally dry material.

In conclusion, it seems clear that there are factors such as structural irregularities and the various molecular and crystallite phenomena observed by Luikov[C] which can cause the sublimation front to be diffuse; however, these appear to operate over short distances. As the melting temperature of the water inside the material being dried is approached, it is likely that the sublimation front becomes more diffuse. However, for freeze-drying conditions which lead to desirable product quality the postulate of a sharp transition from a frozen region to a nearly dry region is useful and more accurate than many other assumptions made in the design and analysis of freeze-drying processes.

DRYING RATES

Pertinent Factors

The factors which interact to govern rates of freeze-drying are indicated schematically in Figure 4. Since freeze-drying necessarily involves sublimation of water, the heat of sublimation (about 1200 Btu/lb of water) must somehow be supplied from a *heat source*. It is also necessary that the generated water vapor be removed by a *moisture sink*. The heat travels from the source to the sublimation front under the impetus of a *temperature difference* driving force (temperature of heat source minus temperature of sublimation front), and the water vapor travels from the sublimation front to the moisture sink under the impetus of a *water vapor partial pressure difference* driving force (partial pressure of water vapor in equilibrium with the sublimation front minus partial pressure of water vapor in equilibrium with the moisture sink). The rates of heat and mass transport in response to a given driving force are determined by the *heat transfer coefficients* and the *mass transfer coefficients*. The heat flux to the material being dried is given by the equation

$$q = \frac{1}{\frac{1}{h_e} + \frac{1}{h_i}} (T_e - T_f) \ . \qquad (1)$$

where

q = heat flux (Btu/hr—ft²)

T_e = temperature of the heat source

T_f = temperature of the sublimation front

h_e = heat transfer coefficient external to the material being dried (Btu/hr—ft²—°F)

h_i = internal heat transfer coefficient, within the material being dried (Btu/hr—ft²—°F).

This same heat flux may also be related to the outer surface temperature of the material being dried, T_s, through the following equation:

$$q = h_i (T_s - T_f) \ . \qquad (2)$$

From Equations 1 and 2 it may be seen that the outer surface temperature at given T_e and T_f depends upon the ratio of the external heat transfer coefficient to the internal heat transfer coefficient:

$$\frac{T_e - T_s}{T_s - T_f} = \frac{h_i}{h_e} \ . \qquad (3)$$

In a similar way, the mass flux of water vapor away from the sublimation front is given by

$$N_A = \frac{1}{\frac{1}{k_{ge}} + \frac{1}{k_{gi}}} (p_{fw} - p_{ew}) . \qquad (4)$$

where

N_A = mass flux of water vaper (lb-moles/hr—ft²)

p_{fw} = partial pressure of water vapor in equilibrium with the sublimation front

p_{ew} = partial pressure of water vapor in equilibrium with the moisture sink

k_{ge} = mass transfer coefficient external to the material being dried

k_{gi} = internal mass transfer coefficient, within the material being dried

If h_e, h_i, k_{ge} and k_{gi} are determined by the

FIGURE 4

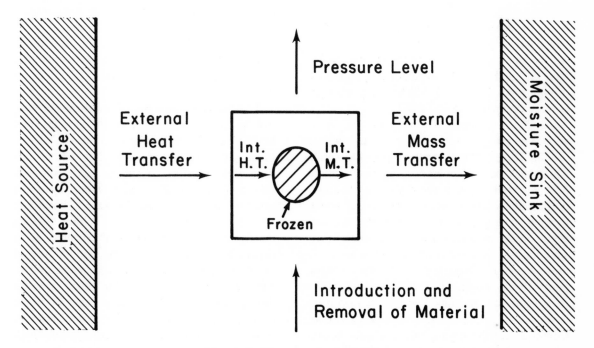

Factors influencing rates of freeze-drying.

natures of the material and of the drying apparatus, and if T_e and P_{fw} are set by the operating conditions, Equations 1 and 4 become two independent equations in four unknowns (q, N_A, T_f and p_{fw}). Two more equations are needed. Unless the sublimation front or piece temperature is changing rapidly, it will be necessary that

$$q = (\Delta H_s)\, N_A \, , \qquad (5)$$

where $\Delta H_s =$ latent heat of sublimation of ice (Btu/lb-mole). The additional necessary relationship is that corresponding to equilibrium at the sublimation front (the equilibrium sublimation pressure of ice):

$$p_{fw} = f(T_f). \qquad (6)$$

Thus the drying rate (N_A) is uniquely determined.

Combining Equations 2, 4, and 5 gives

$$h_i (T_s - T_f) \qquad (7)$$

$$= (\Delta H_s) \left(\frac{1}{\frac{1}{k_{ge}} + \frac{1}{k_{gi}}} \right) (p_{fw} - p_{ew}).$$

From Equation 7 it may be seen that the water vapor partial pressure driving force ($p_{fw} - p_{ew}$) will be small when h_i is small compared to k_{ge} and k_{gi}, and that the temperature difference driving force ($T_s - T_f$) will be small when h_i is large compared to k_{ge} and k_{gi}.

As T_e and, hence, T_s are raised so as to accelerate rates of freeze-drying through higher driving forces, there are two possible limits which may be encountered in a process where heat is supplied across the dry layer. The first is the outer surface temperature, T_s, at which thermal damage, such as denaturation, will occur on the product surface. The second possible limit is the melting of the frozen zone; T_f must be kept adequately below this melting point. In a situation where h_i is low compared to k_{gi} and k_{ge}, the outer surface temperature limit will be encountered first as T_s is raised. Consequently the drying rate can only be accelerated further by somehow increasing h_i,

and, as a result, the process is considered to be *heat-transfer controlled*. Most freeze-drying processes carried out currently are heat-transfer controlled.

As shown in Figure 4, one factor influencing mass and heat transfer coefficients is the total pressure (inerts + water vapor) in the drying chamber. A number of authors[E,25,26,33] have shown that increasing the total pressure tends to decrease k_{gi} (and probably also k_{ge}) and to increase h_i. Consequently, at some level of total pressure h_i increases enough relative to k_{ge} and k_{gi} so that the melting limit on T_f is reached first as the outer surface temperature is raised. At this point the rate of the freeze-drying process has become *mass-transfer controlled* since a higher rate can only be achieved by increasing k_{gi} and/or k_{ge}. The total pressure level at which this transition from heat transfer control to mass transfer control occurs is typically 10 to 20 mm Hg for meats[E,26] but is much lower for substances such as food liquids which contain a high soluble solids content and thus have to be kept below a much lower frozen zone temperature.

Another pertinent point should be noted from Equation 7. As has been pointed out by Strasser,[34] it is not necessary to hold p_{ew} at the lowest possible value during freeze-drying. All that is needed is for p_{ew} to be substantially smaller than p_{fw} in Equation 7, and if p_{ew} is one-tenth or less of p_{fw} there is little to be gained in either rate or product temperature during drying through further reductions in p_{ew}.

The foregoing analysis of heat transfer control vs. mass transfer control assumes that heat is supplied from the outer dried surface. In the freeze-drying of frozen liquid foodstuffs it is often possible to supply heat from one side and remove the water vapor from the other.[9,35] In such a case the rate is limited in all cases by the necessity of avoiding melting the frozen layer adjacent to the heater. T_s in Equation 7 then refers to the temperature of the frozen layer adjacent to the heater and h_i becomes the coefficient for heat transfer through the frozen layer. A process in which heat is supplied through the frozen layer is usually *mass transfer controlled* since the high thermal conductivity of frozen materials makes

h_i large and for most of the drying time T_f is close to T_s.

A final point to be stressed is that in any freeze-drying process it will be desirable to fix the design and set conditions so that the process is not rate limited by external resistances to either heat or mass transfer. The internal heat and mass transfer resistances are characteristic of the material that is being dried, but the external resistances are characteristic of the equipment. When internal resistances to heat or mass transfer are not controlling, it will be possible to accelerate the rate by increasing the external heat transfer coefficient and/or the external mass transfer coefficient, and it will nearly always be profitable in terms of drying capacity to do this. In this way one can achieve the full rate permitted by the internal properties of the material being dried.

External Heat and Mass Transfer Coefficients

The main thrust of freeze-drying research has not been directed toward measurement and interpretation of external heat and mass transfer coefficients in freeze-dryers for two main reasons. First, these external coefficients characterize the equipment more than they do the material being dried, and as a result it is usually expected that standard, already developed approaches for the prediction of heat and mass transfer coefficients can be used. Second, as was pointed out above, in a well designed freeze-dryer with high drying rates the external coefficients should not be controlling in determining the drying time.

Recent Soviet research on rates of sublimation of ice and naphthalene into vacuum[C,13,18,20,30,36-38] have shown the existence of a number of interesting phenomena which increase the external heat and mass transfer coefficients substantially above the values that would be predicted by classical analyses of diffusion, conduction, convection and viscous fluid flow. Among the phenomena reported are rarefaction shock waves and high velocity jets issuing from the subliming surface, caused by the very great volumetric expansion of the vapor generated by sublimation. The volume of water increases by a factor of 10^8 if the chamber pressure is 10

microns. Luikov and co-workers[C,20] have found entrainment of solid crystallites with the outflowing vapor; these crystallites can then serve as nuclei for desublimation or condensation at points in the external boundary layer where supersaturation conditions develop. These effects all serve to accelerate external mass transfer. Brajnikov et al.[18] report measurements of external mass transfer coefficients in a semiindustrial freeze-drying installation which show that the flux due to molecular diffusion under vacuum conditions is only 16% of the total mass flux.

As the pressure in the drying chamber increases, the external mass transfer coefficients should more closely obey the predictions of simple convective and conductive mass transfer theory. Sandall[30] found that observed external mass transfer coefficients for the freeze-drying of single slabs of turkey meat in stagnant helium at 3 to 25 mm Hg agreed closely with the predictions of a simple diffusion analysis, and found that external mass transfer coefficients observed for freeze-drying in circulating helium or nitrogen at higher pressures gave satisfactory agreement with the predictions of laminar boundary layer theory.

Internal Heat and Mass Transfer Coefficients—the Uniformly Retreating Ice Front (URIF) Model[E]

Figure 5 shows the basic model of a uniformly retreating ice front which has been used by most investigators for analyzing internal heat and mass transfer resistances governing rates of freeze-drying. For a process in which heat is supplied to the outer, dry surface, heat transfer is pictured to occur by conduction across the dry layer, from the outer surface to the frozen zone. For one-dimensional drying of a slab, h_i in Equations 1 and 2 then becomes

$$h_i = \frac{k}{\Delta L} , \qquad (8)$$

where

k = thermal conductivity of the dry layer

ΔL = thickness of the dry layer at any time during drying.

Mass transfer of water vapor is pictured to occur in the opposite direction across the dry layer, so that k_{gi} in Equation 4 becomes

$$k_{gi} = \frac{D'}{RT \ \Delta L} \ , \tag{9}$$

where D' is an effective diffusivity of water vapor within the dry layer. R is the gas constant and T is absolute temperature.

Equations 8 and 9 imply that the heat and mass transfer processes occur at a pseudo steady-state across the dry layer, and Equation 8 implies that the outflow of water vapor does not distort the temperature profile greatly from linearity. Sandall et al.[E,30] explored both these assumptions theoretically and ascertained that the maximum effect from deviations from either assumption for freeze-drying of meats was 4% of the predicted rate. Hence, both these effects can be safely ignored in most cases. As was pointed out above, Sandall et al.[E,30] also found a maximum effect of 3%

from neglect of the residual water content of the dry layer if the dry layer achieves local equilibrium with the vapor phase. Gunn[39] found, theoretically, that thermal diffusion makes a negligible contribution to rates of freeze-drying under ordinary conditions, and found, experimentally,[39,40] that surface diffusion also does not make a perceptible contribution to water vapor transport in meats under freeze-drying conditions. Although other more complex solutions for various freeze-drying situations incorporating some or all of these effects have been suggested or published,[15,41-44] the virtues of obtaining a simple, easily used model for freeze-drying rates by neglecting the effects altogether should nearly always outweigh any gain in accuracy from the more complex solutions. This is not to say that more complex analyses are not of value for other types of drying processes.

Slab Geometry—For the one-dimensional slab geometry shown in Figure 5, ΔL may be related to the fraction of the initial moisture remaining, X, by

$$\Delta L = (1-X) \ L/2 \ , \tag{10}$$

under the assumption that all the moisture is removed from a region as the ice front passes through. L is the thickness of the specimen, if drying occurs from both faces.

The rate of freeze-drying can be related to N_A by equating the rate of change of water content to the water vapor flux:

$$\frac{L}{2} \cdot \frac{1}{M_w \ V_w} \cdot \left(-\frac{dX}{d\theta}\right) = N_A \ , \tag{11}$$

where

$\quad M_w$ = molecular weight of water
$\quad V_w$ = volume of the material occupied by unit weight of water initially
$\quad \theta$ = time.

V_w is equal to $(M_o - M_f)/\rho_D$, where

$\quad M_o$ = initial moisture content, lb. water/ lb. dry solid
$\quad M_f$ = moisture content in equilibrium with drying chamber humidity
$\quad \rho_D$ = density of dry solids, including pore spaces (lb/ft^3)

FIGURE 5

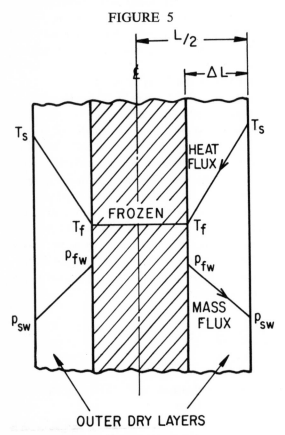

Uniformly retreating ice front model.

Combining Equations 2, 5, 8, 10 and 11 gives an equation relating the drying rate, the thermal driving force and the remaining moisture content:

$$(1-X) = \frac{4kM_w V_w}{\Delta H_s L^2} \cdot \frac{(T_s - T_f)}{\left(-\dfrac{dX}{d\theta}\right)} \quad . \quad (12)$$

A similar equation may be written in terms of the overall temperature driving force, $(T_e - T_f)$, starting from Equation 1.[E]

$$(1-X) = \frac{4kM_w V_w}{\Delta H_s L^2} \left\{ \frac{T_e - T_f}{\left(-\dfrac{dX}{d\theta}\right)} - \frac{2}{L} \frac{k}{h_e} \right\} \quad (13)$$

and a similar derivation[E] relates the drying rate to the overall mass transfer driving force:

$$(1-X)$$
$$= \frac{4D'M_w V_w}{RT\,L^2} \left\{ \frac{p_{fw} - p_{ew}}{\left(-\dfrac{dX}{d\theta}\right)} \right\} \frac{2}{L} \cdot \frac{D'RT}{k_{ge}} \quad (14)$$

If $T_e - T_f$ is assumed constant, Equation 13 may be integrated to give

$$(1-X)$$
$$= \frac{8kM_w V_w}{\Delta H_s L^2} (T_e - T_f) \cdot \left\{ \frac{\theta}{1-X} \right\} - \frac{4}{L} \cdot \frac{k}{h_e} \quad .(15)$$

Equations 12 and 14 may be integrated in similar fashion if $(T_s - T_f)$ or $(p_{fw} - p_{ew})$ is taken constant.[E] Once the temperature of the frozen zone has been determined by equating the rates of heat and mass transfer and by using the relationship for the equilibrium sublimation pressure of ice, any of the last four equations may be used to predict the rate of freeze-drying according to the URIF model.

The time, θ, predicted for completion of drying following the URIF model may be obtained by substituting $X = O$ in one of the integrated equations and solving for θ. If neither $T_e - T_f$,

nor $T_s - T_f$ nor $p_{fw} - p_{ew}$ may be taken as constant during drying, it is necessary to integrate Equations 12, 13 or 14 numerically or graphically from $X = 1$ to $X = 0$ in order to obtain the predicted time for complete drying.

Other Geometries—A solution for the rate of freeze-drying for an isotropic spherically shaped medium has been given by Clark and King[26] as

$$T_e - T_f \quad (16)$$
$$= \left(-\frac{dX}{d\theta}\right) \cdot \frac{r_o \Delta H_s}{3M_w V_w} \cdot \left[\frac{1}{h_e} + \frac{r_o}{k} \left(\frac{1}{X^{1/3}} - 1 \right) \right]$$

for the case where $T_s - T_f$ is constant or where the direct additivity of internal and external heat transfer resistances is assumed. r_o is the sphere radius. A similar mass transfer equation may be written. The equation for a circular cylindrical geometry analogous to Equations 12 and 16 is

$$T_e - T_f$$
$$= \left(-\frac{dX}{d\theta}\right) \cdot \frac{R_o \Delta H_s}{2M_w V_w} \cdot \left[\frac{1}{h_e} + \frac{R_o \ln(X^{-\frac{1}{2}})}{k} \right] (17)$$

where, again, a constant value of $T_s - T_f$ or the direct additivity of internal and external heat transfer resistances is assumed. Again, a similar mass transfer equation may be written. R_o is the cylinder radius.

The foregoing equations apply to the case where heat is supplied from the outer, dry surface of the piece being freeze-dried. When frozen liquid foods (milk, coffee, etc.) are freeze-dried, it is usually possible to secure a good enough contact between a heater plate (or tray bottom) and the frozen region so as to conduct heat in through the frozen region and let vapor escape from the other side of a slab geometry. In this case an equation for heat conduction across the frozen layer replaces or augments the equation for heat conduction across the dry layer, and the right-hand side of Equation 10 must be multiplied by 2, since drying can occur from only one face of the slab. A complication arises from the fact that

T_f will vary with time, making a numerical or graphical integration of the rate equation necessary. Analyses of drying rates when heat is supplied through the frozen region have been given by Harper and Tappel,[9] by Lambert and Marshall[35] and by Dyer and Sunderland.[42]

Many research groups have endeavored to test the applicability of the URIF model by plotting freeze-drying rate data according to the forms indicated by Equations 12 through 16, or by related equations. Ginnette et al.[45] used a model assuming a spherical geometry for heat conduction along with linear mass transport to obtain a linear form of plotting for data for rates of freeze-drying of carrot, apple and beef dice. Kessler[46] tested the URIF model by following temperature profiles during freeze-drying of beds of glass spheres, apple, milk and a hygroscopic building material. The temperature profiles are similar to those reported by Hatcher[27] (Figures 2 and 3), showing a relatively flat temperature gradient within the frozen zone and a rather sudden transition upward in temperature vs. time as the ice front passes a given thermocouple location. Lusk et al.[47] plotted data for freeze-drying of slabs of salmon, haddock and perch in a form indicated by Equation 12 $[dX/d\theta$ vs. $(T_s-T_f)/(1-X)]$ and found a linear relationship over a portion of the drying time. Massey and Sunderland[48] plotted rate data for the freeze-drying of beef slabs according to Equation 12 and obtained linearity over a portion of the drying time, similar to the results obtained by Lusk et al. Massey and Sunderland varied the chamber pressure from 0.2 to 3 torr, and found that the apparent thermal conductivity from the slope of the plot of $dX/d\theta$ vs. $(T_s-T_f)/(1-X)$ increased with increasing pressure, as it should. They also carried out an experiment wherein a slab of ice was sandwiched in between two layers of already freeze-dried beef and sublimation rates were measured. Bralsford[17] interpreted rates of freeze-drying of beef slabs in terms of a URIF model and found linearity for much of the drying time when the data were plotted in a form corresponding to a rearrangement of Equation 15, ignoring the term involving h_e. Gaffney and Stephenson[49] plotted rate data for the freeze-drying of frozen solutions of cornstarch and water as $(T_e-T_f)/$

$(1-X)^2$ vs. θ (Equation 15, ignoring the h_e term) and found linearity and an inferred thermal conductivity which increased with increasing pressure. Magnussen[50] calculated thermal conductivities during the freeze-drying of beef by postulating a pseudo steady state and relating drying rates to measured differences in temperature between thermocouples imbedded in the specimen. The thermal conductivities calculated in this way scattered somewhat during a drying run, but showed no major trend with respect to time. Sharon and Berk[51] plotted rate data for the freeze-drying of tomato juice in the

form of $(1-X)^2$ vs. $\int_O^\theta (T_s-T_f)\ d\theta$, following Equation 15, and found linearity. They also were able to make the corresponding mass transfer plots, which they report were linear.

The eight tests of the URIF model mentioned above were carried out at relatively low drying chamber pressures, such as are typical of most freeze-dryers. Sandall et al.[E, 30] tested the form of the drying curve against the predictions of the URIF model for slabs of turkey meat over a range of pressures (inerts plus water vapor) from 0.15 to 760 mm Hg pressure. At the lower pressures it was possible to plot the rate data as $(1-X)$ vs. $\theta/(1-X)$, following Equation 15, and at the higher pressures the drying rate data were plotted as $(1-X)$ vs. $(p_{fw}-p_{ew})/(-dX/d\theta)$, following Equation 14. In both cases the relationships were linear for the removal of the first 65 to 90% of the initial water present, but the rates slowed thereafter. The results of Sandall et al.[E] show that some of the variations from linearity in plots made in other work can be explained by allowing for the contributions of external heat and mass transfer coefficients or for very slight misplacements of surface thermocouples.

The linearity of the various forms of plotting suggested by the URIF model serves as a test of the ability of that model to fit experimental data for correlations, but in themselves these plots do not confirm that the mechanism of internal heat transfer is solely conduction through a growing dry layer and that the mechanism of mass transfer is combined bulk diffusion, Knudsen diffusion and viscous flow through a growing dry layer. These mechanisms can best be confirmed by comparing transport parameters inferred from drying rate studies with transport parameters measured independently by a different means. Because of the piece-to-piece variability of natural food substances the independent measurement of a transport property should be made on the same specimen with which the drying rate data to be compared were obtained. Studies of this latter sort have been made by Sandall et al.[E] and by Gunn and King.[40]

Sandall et al.[E] used a thermopile apparatus constructed by Triebes and King[32] to measure thermal conductivities of 20 samples of turkey breast meat for which thermal conductivities had been inferred from the linear portion of the $(1 - X)$ vs. $\theta(1 - X)$ plot. The environmental gas and its pressure in the thermopile apparatus were the same as in the chamber during the drying experiment in each case. For the 20 samples the average absolute deviation of the thermal conductivities measured in the two ways was 10.4%, and the thermal conductivity derived from drying rates was, on the average, 8.2% higher than that measured in the thermopile apparatus. This agreement affords a striking confirmation of the URIF model and of a heat transfer by conduction mechanism for the period during which the first 65 to 90% of the initial water is removed. The higher thermal conductivities in the drying experiments may reflect a finite moisture content in the dry layer, but such an inference is probably not warranted by the natural inhomogeneity of the product. Gunn and King[40] used a diffusion cell apparatus to measure rates of counterdiffusion of gases through various of the turkey meat samples freeze-dried by Sandall et al. From the observed diffusion rates, Gunn and King evaluated structural parameters (see below) and thereby independently predicted values of D' for eleven of the specimens freeze-dried by Sandall et al. at higher chamber pressures. The average absolute deviation of D' measured in the diffusion cell and D' derived from the linear portion of the drying curves was 15%, thus, again, lending surprisingly strong confirmation of the URIF model and the proposed water vapor transport mechanism during the removal of the first 65 to 90% of the initial

water. There was no predominant directional deviation of the two measurements of D' from one another for specimens dried in the direction parallel to the meat fibers, but the value of D' derived from the drying curve was uniformly higher than that measured in the diffusion cell for three samples dried in the direction perpendicular to the direction of the fibers.

Bralsford[17] also compared thermal conductivities derived from drying curves for beef with those measured on the dry sample afterwards in a thermopile apparatus for seven samples of beef. He found an average absolute deviation of 5%, with the value derived from the freeze-drying rate averaging 2% higher than the value measured for the dry sample. Magnussen[50] found that thermal conductivities derived from freeze-drying rates for four samples of beef agreed within an average absolute deviation of about 9% with thermal conductivities measured for other samples of beef. These results lend further confirmation to the mechanism of heat transfer by conduction across the dry layer.

Thermal Conductivities

A number of investigators have made thorough measurements of thermal conductivities of samples of various freeze-dried foodstuffs. These thermal conductivities are needed for the prediction and analysis of the internal heat transfer coefficient during freeze-drying (Equation 8). Thermal conductivities of freeze-dried foods as influenced by the pressure and nature of the surrounding gas phase were first systematically investigated by Harper.[52,53] Typical results are shown in Figure 6, for freeze-dried pear. At very low pressures the thermal conductivity reaches a lower asymptotic value which is independent of the surrounding gas. This asymptotic conductivity reflects the geometric structure of the solid matrix itself, with no contribution from the gas in the voids of the material since the gas pressure is so low. At high pressures the thermal conductivity levels out again at a higher asymptotic value. This higher asymptote is characteristic of the heterogeneous matrix composed of solid material and the gas in the voids. Consequently, the high pressure thermal

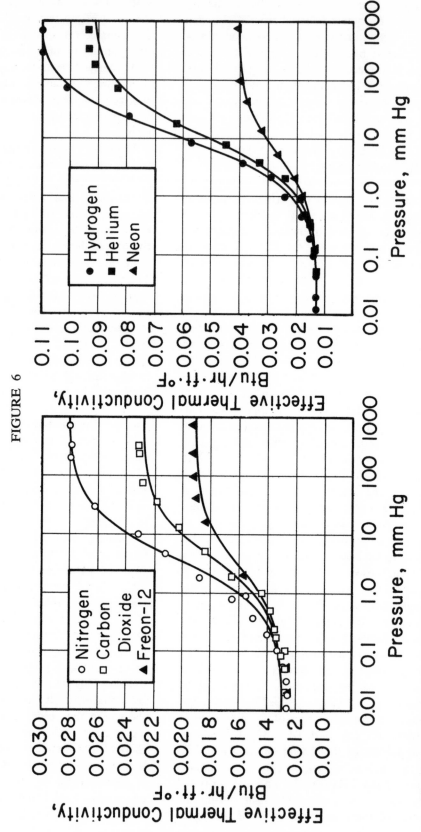

FIGURE 6

Thermal conductivity of freeze-dried pear, as influenced by pressure and nature of surrounding gas, after Harper and El Sahrigi.[53] Reprinted from *Ind. Eng. Chem. Fundam.*, 3, 322, 1964. Copyright 1964 by the American Chemical Society. Reprinted by permission of the copyright owner.

26

conductivity is dependent upon the nature of the gas which is present, and specifically increases as the thermal conductivity of the gas increases and, hence, as the molecular weight of the gas decreases. The high-pressure asymptotic thermal conductivities for the six different gases in freeze-dried pear shown in Figure 6 follow this trend. When the thermal conductivity attains the high-pressure asymptotic value, the mean free path of the gas molecules within the void spaces of the solid, dry foodstuff has become substantially less than the dimensions of the void spaces. During the transition in thermal conductivity from the low-pressure asymptote to the high-pressure asymptote the mean free path of the gas molecules rivals the void space dimensions in magnitude, but once the mean free path is reduced to the point where the gas phase within the solid matrix obeys simple kinetic theory the thermal conductivity stops rising. This reflects the fact that the thermal conductivity of a gas obeying simple kinetic theory is independent of the pressure.

The transition in thermal conductivity between asymptotes in Figure 6 occurs between 0.1 and 100 mm Hg, which includes the pressures characteristic of freeze-drying processes. The pressure range over which the transition in thermal conductivity between asymptotes occurs is characteristic of the pore-size distribution of the void spaces within the freeze-dried material.[31,32,52,53] A smaller pore dimension means that the gas must achieve a higher pressure in order for the mean free path of the gas to become comparable to the pore spacing and, hence, means that the transition between asymptotes will occur at higher pressures. Since fast freezing before freeze-drying leads to smaller pore spacings after freeze-drying (see below), it follows that faster freezing should lead to lower thermal conductivities at a given pressure. If a freeze-drying process is rate limited by internal heat transfer, the rate of freeze-drying for fast-frozen material should then be less than that of a slow-frozen material. Slower freeze-drying rates for food pieces frozen more rapidly have been reported by Karel[54] and by Lusk et al.[55] for shrimp, by Carl and Stephenson[56] for potatoes and mushrooms, and by Haugh et al.[57]

and Hamre[58] for chicken.

The variability of thermal conductivity from one sample to another of the same natural food substance can be quite large. For example, Triebes and King[32] report measurements of thermal conductivity made for 16 different samples of freeze-dried turkey meat at various pressures. Extrapolations to the high pressure asymptote yield a standard deviation of 12% in thermal conductivity; whereas extrapolations to 100 microns pressure yield a 24% standard deviation in thermal conductivity. This can be a cause of uneven drying in freeze-drying processes. Triebes and King[32] also found that the thermal conductivity in the direction of the fibers in freeze-dried turkey meat is substantially greater than that perpendicular to the fiber direction for a number of samples; natural products can often be anisotropic in this way. Yet another finding from Triebes and King[32] and from Saravacos and Pilsworth[59] is that the thermal conductivity of freeze-dried materials is higher at higher relative humidities of the surrounding gas, in rough proportion to the volume fraction of sorbed water present, weighted in proportion to the thermal conductivity of liquid water.

In the freeze-drying of frozen liquid foodstuffs the initial dissolved solids content can have a marked effect on the thermal conductivity of the freeze-dried product. Because of the reduced porosity and increasing contribution of the solid matrix, the thermal conductivity of the freeze-dried product increases as the initial solids content increases. This result has been confirmed for tomato juice of various degrees of preconcentration by Sharon and Berk[51] and for maté extract by Spiess et al.[60]

Various writers have suggested models through which the thermal conductivity resulting from the heterogeneous mixture of the solid matrix and the gas phase in the voids can be predicted and analyzed.[32,46,52,53,88] The curves drawn in Figure 6 are based upon such a model.

Table 1 summarizes thermal conductivity measurements which have been reported by various investigators for different freeze-dried foodstuffs. The surrounding gases and the pressure range are indicated, along with the

TABLE 1
Thermal Conductivities of Freeze-Dried Food Substances

Investigator(s)	Method[a]	Food Substance(s)	Surrounding Gas	Range of Pressures (mm Hg)	Range of Thermal Conductivities (Btu/hr ft °F)
Bralsford[17]	A, B	Beef	Water vapor	0.5–2.4	0.020–0.032
Cowart[61, 49]	A	Mushrooms	Air	0.3–760	0.006–0.021
Gaffney and Stephenson[49]	B	Cornstarch solutions	Water vapor and air	0.1–2.0	0.008–0.019
Harper[52]	A	Beef	Air	0.001–760	0.022–0.038
		Apple	Air	0.001–760	0.009–0.024
		Peach	Air	0.001–760	0.009–0.025
Harper and El Sahrigi[53]	A	Pear	Freon-12 Carbon dioxide Nitrogen Neon Helium Hydrogen	0.02–760	0.013–0.108
		Apple	same	0.02–760	0.013–0.115
		Beef	same	0.02–760	0.022–0.116
Kessler[46]	B	Apple	Water vapor	0.01–0.3	0.020–0.067
		Milk	Water vapor	0.01–0.3	0.013–0.047
Lusk et al.[47]	B	Salmon	Water vapor	0.15	0.024–0.077
		Haddock	Water vapor	0.08	0.011–0.015
		Perch	Water vapor	0.08	0.013–0.020
Magnussen[50]	A, B	Beef	Water vapor and air	0.007–80	0.020–0.037
Massey and Sunderland[48]	B	Beef	Water vapor Air and	0.2–3.0	0.030–0.042
Saravacos and Pilsworth[59]	A	Potato starch	Water vapor	0.03–760	0.005–0.024
		Gelatin	same	0.03–760	0.009–0.024
		Cellulose gum	same	0.03–760	0.011–0.032
		Egg albumin	same	0.03–760	0.008–0.024
		Pectin	same	0.03–760	0.007–0.029
Sharon and Berk[51]	B	Tomato juice	Water vapor	0.4–1.5	0.020–0.100
Sandall et al.;[E] Triebes and King[82]	A, B	Turkey	Air Water vapor Freon-12 Carbon dioxide Helium	0.01–760	0.008–0.112

a A—Thermopile apparatus; B—from drying rate.

range of thermal conductivities encountered. Thermal conductivity measurements inferred from actual freeze-drying rate measurements should be used with some care, because of additional factors which may have been important without being adequately allowed for in the analysis. It will probably be helpful in many cases to make use of the thermal conductivity models for porous media referred to above in order to extrapolate and interpolate data to different conditions.

One very important point brought out by Table 1 is that the thermal conductivities of freeze-dried foodstuffs are quite low; they are excellent insulators. This factor and the limitation placed upon the thermal driving force by the nature of the process are the primary reasons for the slow rates encountered for freeze-drying in practice.

Internal Mass Transport

Mass transport of the escaping water vapor through the dry layer can occur by bulk molecular diffusion through inert gas, by Knudsen diffusion and by viscous flow in response to a gradient in total pressure. As was mentioned previously, surface diffusion has been found not to be a significant contributor for meats,[39,40] and thermal diffusion should not be an important contributor.[39] Even with these simplifications the analysis of the interaction of the remaining mass transfer effects is quite complex. The problem of mass transfer under gradients in composition and total pressure in porous media is common with several other fields of research; for example, mass transfer in porous catalyst pellets. Mason et al.[62] and Gunn and King[63] have presented the equation, allowing for the simultaneous effects of Knudsen diffusion, bulk molecular diffusion and viscous flow, which takes the following form for a binary gas system:[63]

$$N_A = -\frac{c_2 D_{AB}^o K_A P}{(c_2 D_{AB}^o + K_m P) RT} \nabla y_A \quad (18)$$

$$-\left[\frac{K_A(c_2 D_{AB}^o + K_B P)}{c_2 D_{AB}^o + K_m P} + \frac{c_o P}{\mu_m}\right] \frac{y_A}{RT} \nabla P$$

where

N_A = flux of Component A

D_{AB}^o = product $D_{AB}P$ (independent of pressure

D_{AB} = binary diffusivity of the A-B gas pair

K_A and K_B = Knudsen diffusivities of Components A and B, respectively

K_m = $y_A K_B + y_B K_A$

P = total pressure

R = gas constant

T = absolute temperature

y_A and y_B $(= 1-y_A)$ = mole fractions of Components A and B, respectively

μ_m = viscosity of the gas mixture.

The Knudsen diffusivity for any component (i) is given by

$$K_i = c_1 \sqrt{RT/M_i} \,, \quad (19)$$

c_0, c_1 and c_2 in Equations 18 and 19 are structural parameters reflecting the geometry of the porous medium. These parameters are independent of the gases present. c_0 is the viscous permeability constant, with dimensions of length squared. c_1 is the Knudsen permeability constant, with dimensions of length. c_2 is the geometric factor for diffusion, which is dimensionless.

Gunn and King[40,64] have shown that the contribution of viscous flow to the total mass transport is small for the freeze-drying of poultry meat. At low chamber pressures or in the absence of an inert gas the Knudsen diffusion term accounts for most of the flux, and at higher pressures when an inert gas is present the bulk diffusion term becomes rate controlling. It should also be noted that it is at the higher pressures, or for the least porous dried products, or for very low frozen zone temperatures that the internal mass transfer behavior will tend to be rate controlling as opposed to internal heat transfer.

If the pressure of the inert gas all along the dry layer is assumed to be equal to the total pressure outside the piece in the drying chamber (a good approximation for chamber pressures of 5 mm Hg and above for freeze-drying of turkey meat), Gunn and King[40,64] show that Equation 18 may be simplified so that D' in Equation 9 can be given by

$$D' = \frac{c_2 D_{AB}^O K_A}{c_2 D_{AB}^O + K_A P} \quad , \qquad (20)$$

where P is the total pressure in the drying chamber. Equation 20 indicates that $1/D'$ should be linear in the chamber pressure. If the freeze-drying rate is rate-limited by internal mass transport, the reciprocal of the drying rate should increase linearly with chamber pressure, i.e., the drying rate itself should decrease with increasing chamber pressure. Since the freeze-drying process will become internal mass transfer controlled *above* a certain pressure (k increases with increasing pressure, while D' decreases with increasing pressure), the highest rate under mass transfer control will occur *at* the pressure of transition from heat transfer control to mass transfer control, and the attainable drying rate will decrease at higher pressures. Sandall et al.[E] confirmed that the reciprocal of D' obtained from freeze-drying rate measurements at higher pressures for turkey breast meat is indeed linear in chamber pressure. Figure 7 shows their results for drying in nitrogen with two different meat fiber orientations and for drying in helium.

For freeze-drying experiments at lower pressures in the absence of an inert gas the mass transport rate should be related to the gradient in total pressure (= water vapor pressure) within the dried layer as follows:

$$N_A = - \left[K_A + \frac{c_o P}{\mu} \right] \frac{dP}{dz} \quad , \qquad (21)$$

where z is distance in the direction of vapor flow. Equation 21 is a simplification of Equation 18 for this case. Hardin[21] and Luikov[C] have used hypodermic probes to monitor the gradient in total pressure during freeze-drying and have obtained qualitative confirmation of the profiles predicted by Equation 21.

A number of workers in catalysis have proposed pore models by which the structural parameters in an equation such as Equation 18 can be predicted from measurements of nitrogen surface areas, pore size distributions by mercury porosimetry, etc. Spiess et al.[65,87] have proposed a model of this sort for mass transfer during freeze-drying, allowing for diffusion in vapor spaces coupled with diffusion within the solid matrix itself. Pore shapes and sizes were inferred from microscopic examination of freeze-dried specimens of egg white and potato starch solutions with initial dissolved solids contents ranging from 2 to 30%. Specimens of egg white gave tubular ice crystals extending through the sample and gave drying rates agreeing well with the predictions of Knudsen diffusion for the observed pore diameters and shapes when drying was carried out under conditions of internal mass transfer control. Potato starch solutions gave a cellular structure upon freeze-drying, and it was necessary to account for the interruptive effect of cell boundaries on the mass transport process in order to account for observed freeze-drying rates under conditions of internal mass transfer control.

Gunn and King[40,64] report measurements of c_0, c_1 and c_2 for a number of turkey meat specimens. Values of c_0 are much more variable than those of the other two structural parameters, probably because of the great sensitivity of viscous flow to the presence of any larger cracks in the material being dried. Values of c_0 and c_1 can be derived from the permeabilities of beef, peach and apple reported by Harper.[52] c_0 and c_1 are both reduced by fiber swelling resulting from moisture sorption.[40]

For freeze-dried turkey meat Gunn and King[40,64] report values of c_2 ranging from 0.15-0.19 for the direction perpendicular to the fibers to 0.5-0.6 for diffusion in the direction parallel to the fibers. The values of c_2 parallel to the fibers are in qualitative agreement with the known porosities of freeze-dried turkey meat and a low tortuosity resulting from a fairly straight capillary structure in the direction of the fibers.[22] The much lower values of c_2 which are derived from the diffusion data of Harper[52] probably result from an important boundary layer resistance external to

FIGURE 7

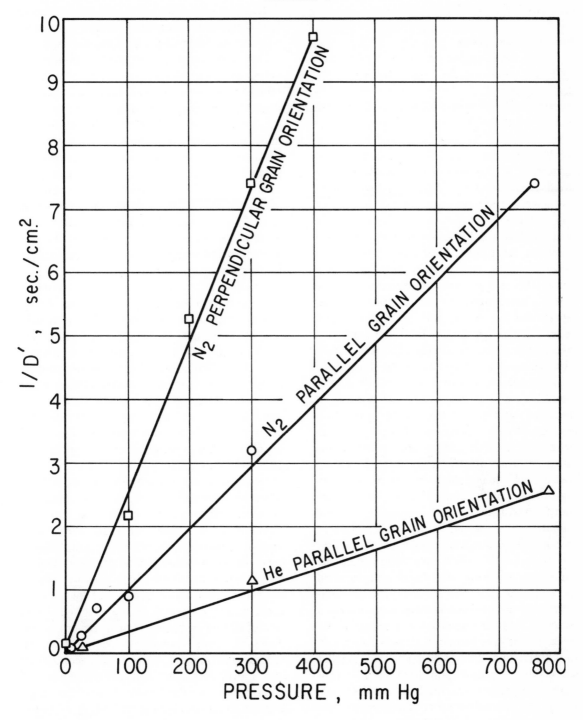

Variation of **D′**, obtained from freeze-drying rates, with respect to pressure, after Sandall et al.[E]

the samples in the device used by Harper to measure diffusion rates.

Frozen Liquid Foods

In the freeze-drying of frozen liquid foodstuffs the mass transfer characteristics of the dry region reflect the initial conditions of freezing and the initial soluble solids content of the substance. Kramers[66] interpreted vapor permeabilities of various dairy products during freeze-drying in terms of the soluble solids contents (e.g., lactose) of the substances being dried. He found that a higher initial solids content gave a denser structure of the dried layer and, hence, a slower rate of freeze-drying which reflected a lower vapor permeability of the outer dried layer. Heat was supplied from a solid surface onto which the sample was frozen; hence heat was supplied through the frozen layer, and the drying process was largely rate limited by mass transfer of water vapor out through the dry layer. This situation is common for freeze-drying of frozen liquids with heat supply across the frozen layer. The lesser permeability with increasing initial soluble solids content reflects the lesser porosity of the dried layer due to voids resulting from ice crystals formed during the freezing process before the eutectic composition is left in the unfrozen concentrate. The eutectic should freeze, itself, during the later part of the freezing process, and it is possible that some mass transfer resistance is also associated with the migration of water out of the small regions of frozen eutectic into the larger voids vacated by ice crystals. Kramers[66] also found that the vapor permeabilities and, hence, the rates of freeze-drying attainable for milk, lactose solutions and orange juice decreased by nearly a factor of 3 as the rate of initial freezing before freeze-drying was increased from 0.2 mm/min to 0.8 mm/min. This trend reflects the smaller ice crystals formed upon freezing at higher rates (see below) and resultant smaller voids left by the removal of ice crystals through sublimation. These smaller voids cause c_1 in Equation 19 to be less since c_1 is proportional to the void dimension. Since the vapor permeation process is for the most part Knudsen flow under the drying conditions commonly used for frozen liquid foodstuffs, a lower c_1 causes a lower freeze-drying rate in direct proportion.

Lambert and Marshall[35] measured rates of freeze-drying of a number of frozen liquid substances, including coffee concentrate, milk and orange juice. Heat was supplied by conduction across the frozen layer, and dry-layer permeabilities could be computed from the observed drying rates. One interesting result was that the initial outer surface temperature of the material at the beginning of freeze-drying assumed a value equivalent to a vapor pressure of water greater than the chamber pressure. From this fact Lambert and Marshall concluded that a dense surface film, or crust, caused an unusually high mass transfer resistance at the surface. They also found that the average permeability of the dry layer changed as drying progressed.

Spiess et al.[60] examined the freeze-drying of maté extract (Paraguayan tea). The drying was carried out at a chamber pressure of 0.58 mm Hg, with heat supplied by radiation from a 90°C source 3 mm removed from a specimen 10 to 15 mm thick. The measured temperatures of the frozen zone during drying reflect the influence of the initial dissolved solids content upon the thermal conductivity and the permeability of the dry layer. As the initial dissolved solids content increased from 2% to 18% the temperature of the frozen zone was observed to increase from −25°C to −18°C, reflecting the combined effects of thermal conductivity and vapor permeability as given by Equation 7. A higher initial dissolved solids content increases k because the solid matrix becomes more dense, but at the same time a higher initial dissolved solids content reduces the vapor permeability because of the reduced porosity and pore size of the dry layer. The rate of drying increased with increasing initial dissolved solids content, indicating that the process was internal heat transfer controlled; but above 20% initial dissolved solids the temperature of the frozen zone rose high enough to cause melting, indicating a transition to internal mass transfer control. Nitrogen surface areas were measured and decreased with increasing initial dissolved solids content because of lower porosity and smaller pore size. Average pore sizes observed from photomicrographs decreased from 400 microns at

1.6% initial dissolved solids to 20 microns at 18% dissolved solids.

Quast and Karel[67] measured vapor permeabilities of 20 and 30 weight per cent coffee solutions and of a model substance containing 10% glucose, 10% microcrystalline cellulose and 2% potato starch. Permeabilities to water vapor were measured in an independent steady-state apparatus and also by graphical differentiation of observed freeze-drying weight loss curves. The URIF model was invoked for the freeze-drying calculations. During freeze-drying, heat was supplied by conduction through the frozen layer, and the rate was assumed to be completely controlled by internal mass transfer resistance (i.e., T_f assumed equal to the temperature of the frozen zone in contact with the heating plate). Samples were frozen under five different conditions, ranging from very slow (free convection at $-5°C$) to very fast (conduction from liquid nitrogen). A primary finding was that most methods of freezing caused the formation of a thin and relatively impermeable surface layer which impeded drying, as was also found by Lambert and Marshall.[35] This surface layer, when sampled in the frozen state, was found by Quast and Karel to contain 35% dissolved solids for the case of coffee which contained 20% dissolved solids in the bulk. The surface layer was much less impermeable for "slush" freezing, in which the liquid was first partially frozen at $-20°C$ with intermittent agitation. To complete freezing, the resulting slush was then solidified at $-40°C$.

Except for the fastest rate of freezing, the water vapor permeabilities measured by Quast and Karel[67] in the steady-state apparatus decrease substantially with increasing pressure; in fact, it appears that the reported permeabilities decrease with very nearly the -1 power of pressure. This result strongly suggests the presence of inert gases in their permeability apparatus, from degassing, leakage or some other source, since binary bulk gas diffusivities are inversely proportional to the total pressure. For samples frozen at the fastest rates prior to drying the reported permeability is either insensitive to pressure or increases with increasing pressure; the lower reported permeabilities and the small pore sizes to be expected for the fastest rate of freezing could well produce a transition to mass transport by Knudsen diffusion. The permeability reported by Quast and Karel does decrease with increased freezing rate before drying, indicating reductions in c_2 and c_1 in Equations 18 and 19 for faster freezing. Because of the apparent presence of inert gases in the steady-state experiments (and perhaps the freeze-drying experiments, as well), because of the variable per cent wise contribution of the impermeable surface layer during the course of the freeze-drying experiments as compared to the steady-state experiments, and because of the demonstrated inapplicability of the URIF model for some of the freezing conditions (see below), it is probably not appropriate to compare the steady-state permeabilities of Quast and Karel with those inferred from the freeze-drying curves.

For their freeze-drying runs Quast and Karel[67] assumed a negligible temperature drop across the frozen zone in order to obtain a partial pressure driving force for the calculation of permeabilities. This temperature drop has an important effect upon the calculated permeabilities, however. Following the calculational procedure given by Quast and Karel in their Figure 5, one finds that the change in temperature of the sublimation interface changes the water vapor partial pressure driving force for mass transfer far enough to substantially increase the permeabilities for the earlier portions of drying. For example, the permeabilities reported for slush freezing in their Figure 8 are raised to a rather uniform level of 2.4 to 3.0 x 10^{-5} g/sec cm mm Hg for thicknesses of 1 to 6 mm of the dry layer, whereas the permeabilities calculated by Quast and Karel, ignoring the temperature gradient across the frozen zone, appear to decrease markedly as dry lever thickness decreases across that range.

Quast and Karel[67] also examined the uniformity of the retreat of the frozen zone for freeze-drying after various preliminary freezing conditions. They found that slush-freezing gave a very uniform retreat, but that other freezing methods gave a nonuniform ice front retreat, which Quast and Karel interpret in terms of the nonuniformity of freezing conditions within the specimens before drying.

Sharon and Berk[51] also found that permeabilities of the dry layer calculated from data on the freeze-drying of tomato juice and tomato concentrates decreased with increasing initial dissolved solids content. Two freezing rates were employed, one being slow, in a $-25°C$ chest for 12 hours, and the other being faster, an air blast at $-30°C$ for one hour. The drying was limited by the outer surface temperature (internal heat transfer control) for the slow-frozen material, but was limited by melting of the frozen zone (internal mass transfer control) for the fast-frozen material. Sharon and Berk confirmed that the mass transfer limitation came from a glazed, relatively impermeable surface layer formed in the blast freezer, by showing that the process returned to heat transfer control if a gauze was frozen into the surface and the surface layer removed along with the gauze.

Monzini and Maltini[68] avoided an impermeable surface layer during the freeze-drying of orange juice by granulating the product after freezing and before drying. This procedure has also been successful for coffee. Granulation of the orange juice was found necessary by Monzini and Maltini for preconcentrated orange juice in order to avoid melting when the surface temperature was held at $35°C$ in a pilot-scale freeze-dryer. The freeze-drying rate was found to increase with increasing initial dissolved solids content, as was also found by Spiess et al.[60] for maté extract, because of the increase in thermal conductivity (internal heat transfer control).

Thijssen and Rulkens[F] froze slabs of dextrin-water model solution as layers of uniform thickness on a plate, with unidirectional freezing. The freezing rate was varied over a wide range. The nitrogen permeabilities of these slabs at atmospheric pressure were measured after the samples had been freeze-dried. The measured permeabilities decreased uniformly with increasing freezing rate before drying. Effective pore diameters calculated by applying a straight, parallel capillary model to the measured permeabilities ranged from 4 microns down to 1 micron. Permeabilities and effective pore diameters were also calculated from measured rates of freeze-drying and measured temperature gradients across the

slabs during drying, with the heat supplied through the frozen layer. The permeabilities measured in the two ways agreed well with one another for 23 samples studied. The rate of freeze-drying was increased for slower initial freezing rates in this case because the process was rate-limited by mass transfer across the dry layer. As noted above, the same trend with freezing rate occurs when the drying rate is limited by heat transfer across the dry layer.

MacKenzie[28,29] has made several observations on the water vapor transport mechanism in the freeze-drying of frozen liquid materials, based upon his studies with the "freeze-drying microscope." He finds that under some conditions it may be necessary to allow for a multiple-series path involving vapor phase transport interspersed with liquid or solid phase diffusion steps, as is pictured in the aforementioned mass transport model of Spiess and co-workers.[65,87]

Terminal Drying Rates

A number of investigators working with freeze-drying of solid foodstuffs have found that the removal of the last 10 to 35% of the initial water content occurs at a rate substantially slower than predicted by the URIF model.[E,45–48] Slow terminal drying has often been attributed to the removal of sorbed, or bound, water.[6,10,16] If equilibrium is reached locally between the dry layer and the vapor phase, the amount of sorbed moisture remaining for heat supply through the dry layer should be small[E] so that any slowing of the rate for bound water removal must reflect a transport limitation of some sort. Rates of bound water removal by diffusion in open pore spaces should be comparable to those already built into the URIF model, since there is not a great amount of shrinkage of the dry layer in freeze-drying.[69] Hence, any added rate limitation for the removal of bound water must lie in the step where the moisture migrates from the interior of a fiber or cell to an open vapor channel. Resistances to moisture migration within the fibers of freeze-dried turkey breast meat were studied by Margaritis and King[22] in the experiment, described above, where transient responses of pressure drop across a sample were monitored after step changes in the humidity of a permeating nitrogen stream. Moisture equilibration was reached at rates corresponding to time constants less than one hour; therefore the resistance to moisture migration within the fibers of turkey meat is not expected to be a significant rate-governing effect during freeze-drying.

Nonuniform retreat of a still-sharp frozen front is another factor which can cause a slowing of terminal rates of freeze-drying, since the early disappearance of the ice from one section of a specimen while ice remains in another zone will cause an unexpected loss of frontal area of the frozen zone for heat transfer and vapor release through sublimation. Margaritis and King[22] report experiments made by Gandhi wherein partially freeze-dried slabs of turkey meat were cut into 16 sections each, perpendicular to the main ice front. The residual moisture content of each of these sections was measured, and appreciable scatter of residual moisture contents from region to region was found. The discolorations observed from melting of the residual water indicated that the ice front when drying was interrupted was relatively sharp, but the scatter from region to region indicates nonuniform retreat of the frozen front. In analyzing the departures from the predictions of the URIF model in the freeze-drying measurements of slabs of turkey meat by Sandall et al.,[E] Margaritis and King conclude that the major factor causing nonuniform retreat and slower terminal rates for runs made with a circulating external gas was the change in external mass transfer coefficient with respect to distance away from the leading edge of the slab. Some drying from the edges of the slabs occurred, despite taping and gluing, and was another cause of nonuniform retreat and slower terminal rates. Other factors which can be important and which can cause nonuniform retreat of the sublimation front within a single piece during freeze-drying are local variations in the effective diffusivity, D', within the piece and local variations in the thermal conductivity within the specimen or the heat transfer coefficient or the intensity of heat supply external to the piece. These latter heat transfer and heat supply factors can cause nonuniform retreat of the ice front only to the extent that the high thermal conductivity of the frozen zone is not able to even out temperatures across the frozen zone. Nonetheless, under conditions of heat transfer control a small aberration in temperature at some point along the sublimation front can have a major impact on the local rate of sublimation because of the great sensitivity of the small mass transfer driving force ($p_{fw} - p_{ew}$) to changes in p_{fw} through changes in temperature (T_f).

When frozen liquid foods are freeze-dried in slab form with heat supply through the ice layer, the residual moisture content corresponding to local equilibrium with the gas phase along the dry layer will be greater because of the relatively lower temperatures within the dry layer. A significant terminal drying period may often be required to remove this moisture.

INFLUENCE OF FREEZING CONDITIONS

Influence of the freezing conditions before drying upon the transport properties during drying have been discussed in the previous sections. In summary, these influences come from a reduction of pore size within the freeze-dried material as the freezing rate before drying is increased, and from the tendency for a relatively impermeable surface layer to result from the freezing methods in at least some cases for liquid foodstuffs. These changes in transport properties affect drying rates and the temperature of the frozen zone during drying. There are also a number of other important influences of freezing conditions upon freeze-drying which have been found in recent years. Some of these are summarized in this and the following sections.

For the freezing and freeze-drying of meats the ice crystal location and the resultant void location in the freeze-dried product are governed by the rate of freezing. Thirty years ago Koonz and Ramsbottom[70] found that the histology of freeze-dried chicken meat is markedly altered by the freezing conditions before drying. Thin slices of chicken frozen in Dry

Ice-acetone at 176°C exhibited ice crystals and voids after drying which were scattered within the individual muscle fibers, many voids to a fiber. Whole chickens frozen in air at −49°C exhibited ice crystals and resultant voids located at the periphery around individual muscle fibers but still contained within the sarcolemma surrounding the fiber. Whole chickens frozen in air at −40°C exhibited single ice columns and resultant voids extending up the center of each fiber, one column to a fiber. Whole chickens frozen in air at −36°C showed some ice crystals and voids at the periphery of fibers within the sarcolemma but also some crystals located outside the sarcolemma and in between adjacent fibers. When chicken was frozen in air at −26°C, the ice formation was almost totally outside the sarcolemma and between adjacent fibers, and at −13° the crystals between fibers left larger voids. These transitions in the ice crystal and resulting void location as the freezing rate becomes slower can be interpreted in terms of the greater ratio of ice crystal nucleation rate to ice crystal growth rate for freezing at lower temperatures, affording less opportunity for water to migrate to a growing crystallite.[70,71] Water transfer across the sarcolemma to ice crystals growing between adjacent fibers occurs by exosmosis[71] and requires the longer times associated with slower freezing rates. Kuprianoff[71] shows photographs obtained by Kallert for frozen beef muscle, which indicate crystal locations at different freezing conditions similar to those described by Koonz and Ramsbottom. Luyet[72] reports similar results from photomicrographs of frozen beef. Margaritis and King[22] have found that the configuration of a single ice crystal extending along the central axis of each fiber is the pattern resulting from common freezing conditions used for freeze-drying. Rapatz and Luyet[73] have found that the structure of the original, unfrozen meat fiber can be closely reattained if a specimen is thawed soon after freezing under conditions causing crystal growth outside the fiber.

King et al.[31] report pore-size distributions measured by mercury intrusion porosimetry for portions of samples of turkey meat frozen at different rates, ranging from 0.6 mm/min to 9.5 mm/min. The results are shown in Figure 8. The average pore sizes inferred from Figure 8 agree qualitatively with the pore sizes obtained by microscopy and reported by Luyet[72] for beef frozen at different rates. King et al. also measured nitrogen surface areas for other portions of the same samples. These vary from 0.3 to 5.3 m^2/g, and calculations[74] show that the nitrogen surface areas agree within a factor of two with the pore-size distributions from mercury intrusion porosimetry reported in Figure 8 if a pore model of non-necking, nonintersecting capillaries is adopted.

In the case of freeze-drying of frozen liquid foodstuffs (fruit juices, coffee, etc.) the nature of the freezing conditions before freeze-drying is critical to the success of the drying process and influences the temperatures under which the freeze-drying may be carried out. Complex factors are involved in an understanding of the freezing of food liquids and of their subsequent behavior during freeze-drying. The most relevant developments along these lines have come from Rey and co-workers and from Luyet, MacKenzie and co-workers. Somewhat arbitrarily, the paper by Rey and Bastien[D] has been picked from among these works as the focal point for review since it touches upon all aspects of the problem and deals specifically with the freeze-drying of orange juice.

Rey and Bastien[D] applied the technique of differential thermal analysis to frozen orange juice and also followed the nature of the frozen state by means of electrical resistivity measurements. They report measurements made for Earle's balanced salt solution with 10% glycine and 5 to 25% glucose added, as well as for orange juice. Rey[75] had earlier applied these techniques to other aqueous solutions. Figure 9 shows the differential thermal analysis curve reported by Rey and Bastien for frozen orange juice upon rewarming, and Figure 10 shows the corresponding electrical resistivity curve as a function of temperature. Differential thermal analysis compares temperatures of an unknown and a control heated at the same rate, with the result that heat effects from phase change and any unusual sensible heat effects show up as peaks, i.e., as deviations from a zero temperature difference between the unknown and the control. In Figure 9 the

FIGURE 8

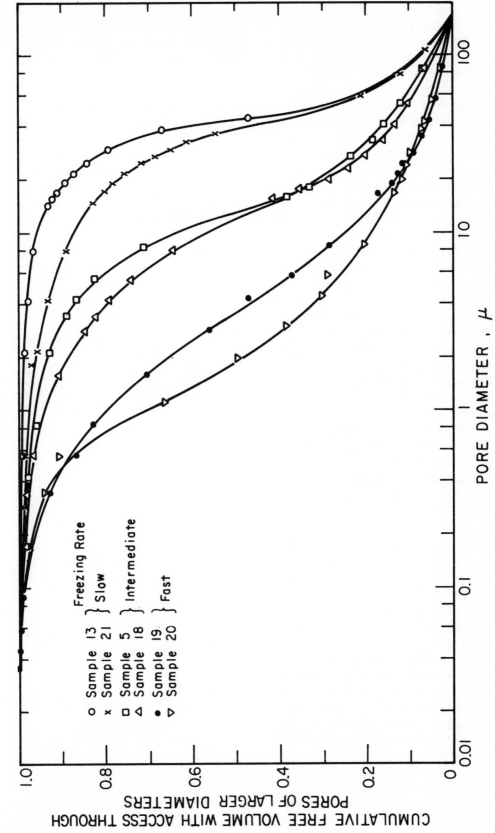

Pore size distributions measured by mercury intrusion porosimetry for freeze-dried turkey meat, after King, Lam and Sandall.[31] Copyright c 1968 by Institute of Food Technologists.

37

differential thermal analysis diagram has been obtained for the rewarming of frozen juice after prechilling to very low temperatures. A melting curve is preferable to a freezing curve for analysis since the melting curve should circumvent subcooling phenomena. In Figure 9 ΔT greater than zero corresponds to the evolution of more heat by the unknown than by the reference, and ΔT less than zero corresponds to the consumption of more heat by the unknown.

In the differential thermal analysis curve of Figure 9 it may be seen that melting of the frozen juice begins to occur at $-32°C$ and continues as the temperature continues to increase toward $0°C$. There is a small exothermic phenomenon indicated by an arrow at $-42.5°C$. The electrical resistivity, shown in Figure 10, decreases continuously as the temperature is raised above $-50°$, indicating progressively more formation of liquid. The resistivity obtained upon cooling from the liquid state down into the frozen region is less at a given temperature than that found upon heating a frozen sample starting from a very low temperature; the high resistances corresponding to a high degree of solidification are reached at lower temperatures upon freezing than occur for the melting curve.[6,75] This is an additional indication of subcooling of the liquid when the temperature is lowered.

Rey[D,75] attributes the heat release at $-42.5°C$ obtained upon rewarming to a phenomenon known as *devitrification* or *recrystallization*. A much more pronounced heat release effect of this sort has been found at low temperatures by Rey[75] for other solutions (e.g., Earle's balanced salt solution $+50\%$ glycerol). Devitrification results from the fact that much of the water present, particularly in a carbohydrate solution, does not crystallize upon initial cooling to temperatures well below the supposed eutectic freezing point, even though the cooling is carried out quite slowly. A sufficiently dilute solution will generate ice crystals as the initial freezing point is surpassed, but crystallization will not continue to occur as the temperature is lowered further. The solution instead becomes extremely viscous at some combination of solute concentration and temperature and becomes vitrified as

a glassy material, without the release of heat of crystallization. Lusena[76] and others have found that subcooled viscous solutions can remain in a nonequilibrium phase state for

FIGURE 9

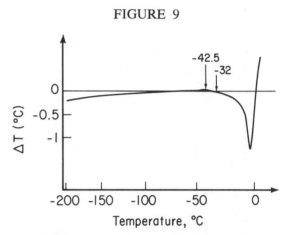

Differential thermal analysis of orange juice, after Rey and Bastien.[D] Reference solution = distilled water. FREEZE-DRYING OF FOODS, Quartermaster Food and Container Institution for the Armed Forces, Division of Engineering and Industrial Research, edited by Frank R. Fisher, National Academy of Sciences—National Research Council, Washington, D. C., 1962.

FIGURE 10

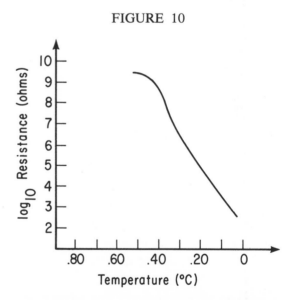

Electrical resistance of orange juice (relative) at low temperatures, after Rey and Bastien.[D] FREEZE-DRYING OF FOODS, Quartermaster Food and Container Institution for the Armed Forces, Division of Engineering and Industrial Research, edited by Frank R. Fisher, National Academy of Sciences—National Research Council, Washington, D. C., 1962.

extremely long periods of time. When the substance containing this glassy material is reheated to a certain temperature (the devitrification or recrystallization point) there is a release of heat, which corresponds to the crystallization of some of the glassy material.

Devitrification has been extensively studied by Luyet and co-workers.[28,29,77-79] The recrystallization temperature can often be 20 or 30°C below the equilibrium melting temperature, this deviation being most pronounced for carbohydrates and glycols. Recrystallization temperatures reported by Luyet,[79] MacKenzie,[29] and Rey[6] are given in Table 2 for

TABLE 2

Recrystallization Temperatures
(After MacKenzie[29] and Rey[6])

Substance in aqueous solution	Recrystallization temperature, °C
Dextran	−10
Fructose	−48
Gelatin	−11
Glucose	−41
Raffinose	−27
Sucrose	−32
Strawberries	−32 to −20
Orange juice	−43 to −18
Apple juice	−43 to −23
Grape juice (Muscat)	−45 to −27
Lemon juice	−22 to −19
Peach juice	−21 to −18
Apricot juice	−29 to −22

various aqueous solutions. Luyet[79] has noted that the recrystallization temperature is insensitive to the initial concentration since it refers to the concentrates formed by removal of some of the water present as ice.

It has frequently been noted that fruit juices and certain other liquid foods are difficult to keep in the fully frozen state during drying. For example, Quast and Karel[67] report that "20% surcrose solutions could not be freeze-dried at −20°C, which is well below the eutectic temperature." They also encountered melting spots during attempted freeze-drying of 40% dissolved solids coffee concentrate at −20°C. Malecki et al.[80] found that apple

juice would not freeze-dry satisfactorily at temperatures above −34°C in a fluidized bed. Poor freeze-drying of these products generally manifests itself as a collapse of the dry layer, or as puffing, foaming, stickiness, or other signs of partial melting. The temperature at which the dry layer will collapse upon freeze-drying is related to the recrystallization temperature, the two being quite close to one another, as has been shown by MacKenzie.[29] Probably the two phenomena reflect the same basic cause.

Rey and Bastien[D] and Rey in earlier experiments endeavored to maintain satisfactory solidification of orange juice and of Earle's balanced salt solution during freeze-drying by monitoring the reading of an electrical resistance probe frozen into the material and adjusting the intensity of heat input so as to maintain a constant resistance reading. Some success was obtained through this method, but the product orange juice was hygroscopic with poor keeping qualities. Much more success was obtained by Rey and Bastien[D] in freeze-drying orange juice which had been given a thermal pretreatment in which the material was first frozen at −40°C, then postfrozen in liquid nitrogen down to −196°C, and then rewarmed to −65°C before being placed into the freeze-drying chamber. The product was less hygroscopic and the drying rate was twice as fast even though the resistance indicated by the probe was held at the same value as in the case of drying without the thermal pretreatment. This fact indicates that the orange juice had achieved some devitrification during the pretreatment and that the frozen zone temperature corresponding to a given resistivity was therefore higher. Heating to the recrystallization temperature and then recooling before starting freeze-drying should, therefore, be helpful in securing a more crystallized frozen material and thereby satisfactory freeze-drying at a higher temperature. Pretreatment to temperatures below the recrystallization temperature may also be of use, since the resistivity measurements by Rey[D,75] indicate that changes of resistivity also occur at temperatures below the apparent recrystallization temperature.

In the case of coffee the color of the freeze-dried product is apparently strongly influenced by the freezing conditions used before drying. Fast freezing leads to loss of the desired dark color of the product; in fact, quite slow freezing is necessary to preserve the color to the extent desired as is shown, for example, by Cruz Picallo.[81] One processing approach which has been suggested for the attainment of a more rapid freezing rate while maintaining a good color retention is a two-step freezing process,[82] similar to the slush freezing used by Quast and Karel[67] and described earlier. Another suggestion has been the use of controlled, intermittent, partial melting and refreezing of the material during freeze-drying so as to retain color without such a slow freezing step being required beforehand.[83]

QUALITY FACTORS AS INFLUENCED BY PROCESSING CONDITIONS

Aroma Retention

One of the major benefits of freeze-drying for many foodstuffs is the relatively good retention of volatile flavor and aroma components with the process. Other methods of dehydration are as a rule inferior to freeze-drying in this respect. One of the first quantitative experiments related to aroma retention in freeze-drying was reported by Rey and Bastien,[D] who added approximately 0.1% acetone to the solutions of Earle's balanced salt solution and glucose described above. They found that the acetone retention during freeze-drying was strongly influenced by the initial glucose content of the solution, increasing from 5% acetone retention when the initial solution contained 5% glucose to 45% retention when the initial solution contained 25% glucose. Furthermore, the acetone retention of the specimen was not reduced during more than 30 hours in a vacuum chamber at 36°C and less than 0.001 mm Hg pressure, subsequent to freeze-drying. This surprisingly good retention was attributed by Rey and Bastien to an adsorption phenomenon; however, it is not clear from measured adsorption isotherms of organic compounds upon freeze-dried substances, such as those reported by Issenberg et al.,[84] how these compounds could be held onto adsorption sites more strongly than water, which is highly polar and strongly adsorbed.

Improved aroma retention at higher contents of dissolved solids was confirmed by the controlled experimental studies made by Saravacos and Moyer[85] of the retention of added compounds (ethyl acetate, methyl anthranilate, acetic acid and ethyl butyrate) during the freeze-drying of apple slices and pectin gels. The rate of loss of these compounds became less as the residual solids content became greater during the course of drying. Chandrasekaran and King[86] measured the retention of various naturally occurring aroma compounds in apple juice during sublimation and/or evaporation of a portion of the water from apple juice which was frozen to different extents. Flame ionization chromatography and vapor headspace analyses were used to monitor the retention of the naturally occurring compounds. Chandrasekaran and King found that the retention of aroma substances during drying from a highly frozen state was considerably superior to the retention observed for evaporation from the totally liquid state, but they also found that the aroma retention during drying from a partially frozen state, corresponding to a temperature of about −16°C, gave aroma retentions nearly the same as those obtained in "freeze-drying" experiments where the temperature was below −25°C. They called drying from this partially frozen state "slush" drying. In both freeze-drying and slush-drying the volatiles retention during a given percentage water removal improved significantly with increasing initial degrees of concentration of the apple juice. No such trend was apparent for drying from the totally liquid state. To a good approximation the retention of different aroma components during freeze-drying and slush-drying was independent of their volatility in solution.

Thijssen and Rulkens[F] present an analysis

of the aroma loss problem for freeze-drying of frozen liquid foodstuffs. Their analysis is an extension of their work on the general problem of aroma retention during all forms of drying, which has been presented elsewhere.[89,90] The loss of aroma compounds during drying is treated as a diffusion process. Experimental measurements of diffusion coefficients at ambient temperatures reported by Thijssen and Rulkens[F,89,90] and by Menting[91] show that the binary diffusivity of sugar-water systems decreases substantially with increasing sugar content (wt %) in solution. The diffusion coefficients for organic model aroma compounds in sugar-water solutions decrease much more rapidly with increasing sugar content in solution than do the binary sugar-water diffusion coefficients. Thijssen and Rulkens[F,89] report diffusion coefficients for water in coffee extract which decrease from about 10^{-5} $cm^2/$ sec in dilute solutions in water at ambient temperatures to about 10^{-10} cm^2/sec in "solutions" containing 95 wt % coffee solids in water. Such a high solids content may in fact represent a non-equilibrium liquid state. For acetone present at small concentrations in solutions of coffee solids in water Thijssen and Rulkens[F,89] report diffusion coefficients decreasing from 10^{-5} to 10^{-13} cm^2/sec over the same range of solids contents. The result is that the ratio of the effective diffusion coefficient for acetone to that for water decreases from the order of unity at low solids contents to the order of 10^{-3} at 95% solids content. Similar results are reported by Menting[91] for the diffusion of water and acetone in solutions of maltodextrin in water. The diffusivities reported by Thijssen and Rulkens and by Menting are shown in Figure 11.

This great reduction in acetone diffusion coefficient relative to the water diffusion coefficient is taken by Thijssen and co-workers to be evidence for very selective retention of aroma compounds when water is removed in drying by diffusion through regions of high solids content. Since the highest solids content within a substance being dried will occur near the surface of water evaporation or sublimation, it follows from this analysis that aroma retention will be effected primarily by selective diffusion of water as opposed to organic aroma

compounds in the region of highest solids content near the surface of evaporation or sublimation. Rulkens and Thijssen[92] have presented an approach for numerical solution of concentration profiles for dissolved solids and for aroma substances in the material being dried.

Chandrasekaran[93] has measured diffusion coefficients for aroma species and water in sugar solutions using the diaphragm cell technique, and has confirmed that the effective diffusion coefficients for aroma components in solutions of simple sugars and in apple juice solutions decrease more rapidly with increasing solids content than do the diffusion coefficients for the water-solids binary system. It appears that selective diffusion of water relative to aroma species begins to occur at lower wt per cent solids contents for simple sugars (levulose, glucose, sucrose) than is reported by Thijssen and Rulkens for solutions of higher molecular weight sugars.

The selective diffusion concept of Thijssen and co-workers is more satisfying than the selective adsorption concept for explaining the retention of aroma compounds during freeze-drying. The selective diffusion theory also has the capability of explaining the significantly high aroma retentions which have been observed for some other forms of drying from the unfrozen state.[89,90] With regard to quantitative application of the selective diffusion theory, it should be realized that the aroma retention characteristics predicted by the theory are dependent upon the values of the relatively low diffusion constants found at high solids contents. These diffusion coefficients are quite difficult to measure since they require prohibitively long experiment times in pseudo steady-state measurements such as with diaphragm cells and since the measured values are highly sensitive to small mass fluxes of other components and to transients in operating conditions in dynamic absorption or desorption experiments of the type carried out and described by Menting.[91] However, even when it is invoked qualitatively, the selective diffusion concept is quite useful in analyzing tendences for aroma loss.

Using the selective diffusion concept, Thijssen and Rulkens[F] have analyzed the retention of volatile species during freeze-

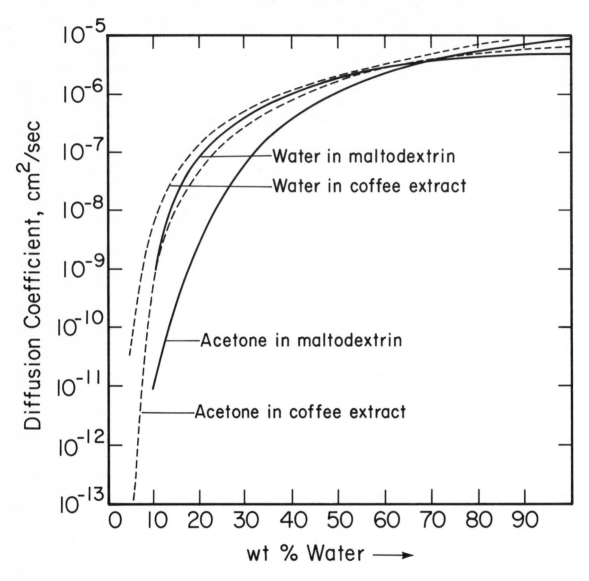

a. Diffusion coefficients of water and acetone in coffee extract and maltodextrin solutions.

drying as influenced by the freezing rate, the dissolved solids concentration, the absolute pressure, the particle dimensions, the means of heating and the dehydration rate. For making this analysis they employed a version of the URIF model in which the ice is initially present in the form of equally spaced, tubular, circular crystals of uniform diameter extending all the way through the specimen. As freeze-drying progresses and water vapor is removed through sublimation, the boundary between ice

and vapor retreats down the column leaving behind a honeycomb of circular pores within a matrix of eutectic solid or highly viscous, immobilized liquid. Before freeze-drying, the aroma compounds are located entirely within this matrix. After the retreat of the ice front past a particular location the residual water and the aroma material have the opportunity to diffuse out through the solid or liquid matrix into the open pores and thereby escape. Also there is an opportunity for aroma compounds to diffuse from the region just within the frozen zone under the ice front up to the region above the ice front and out into the open pores.

By applying dimensional reasoning to the transport processes involved, Thijssen and Rulkens conclude that the degree of aroma retention should increase (1) with increasing drying rate, (2) with increasing thickness of the eutectic solid or immobilized liquid layers between the pores, and (3) with a decrease in water concentration in the eutectic solid or immobilized liquid at the ice front. Although there may be an effect of the dissolved solids content upon the original ice crystal diameter and, hence, indirectly upon the thickness of the eutectic solid or immobilized liquid layer, the primary effect of increasing initial dissolved solids content should be to increase the volumetric ratio of eutectic solid or immobilized liquid to ice crystals before freeze-drying and, hence, to increase the thickness of the layers between open pores. Therefore, increasing the initial solids content should increase

FIGURE 11b

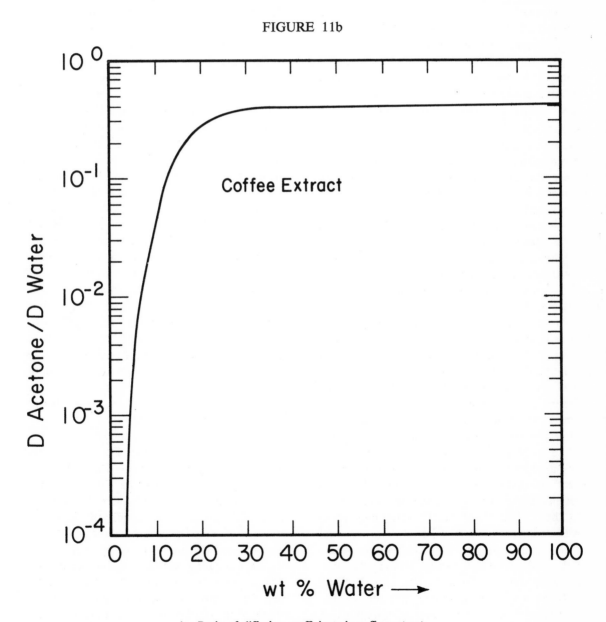

b. Ratio of diffusion coefficients in coffee extract.

43

the degree of aroma retention, a fact which has been observed experimentally. The freezing rate before freeze-drying influences the ice crystal diameter and, hence, the thickness of the layers between pores. Specifically, lower freezing rates cause larger ice crystal diameters and, hence, greater thicknesses of the eutectic solid or immobilized liquid layers. Consequently, slower freezing rates should cause improved aroma retentions upon drying. Thijssen and Rulkens also predict that the influence of ice front temperature upon aroma retention will be manifested as opposing effects of decreased drying rate and decreased water content of the immobilized liquid layer in the case where eutectic material is not present. Except for the effect of increased initial solids content in improving aroma retention, these predictions have not been tested by quantitative experimental data reported in the literature.

Additional insight into the retention of aroma compounds during freeze-drying is given by the work of Karel and Flink,[A,94] who measured the retention of model aroma substances during freeze-drying and also during controlled rehumidification of various carbohydrate solutions. The carbohydrates employed were glucose, sorbitol, maltose, sucrose and 10,000-molecular-weight dextran. The aroma substances were a number of normal alcohols, two branched alcohols, acetone and methyl acetate. The solids content before drying was 20% by weight and the organic volatile component was present to 0.8% by weight, at a level of 4 g/100 g of solids. Measurements of volatiles loss after freeze-drying in an evacuated flask for varying lengths of time[94] confirmed that the most rapid volatiles loss occurs during the early portion of drying and that after a certain point (residual water contents of 15 to 35 g/100 g solids) there ceased to be any appreciable loss of organic volatile component as drying proceeded onward. Volatiles retention after drying ranged from very low up to about 75% of the initial volatile content. It was confirmed that at relatively low residual moisture contents the remaining volatile material could not be removed by prolonged evacuation, as was also found by Rey and Bastien.[D]

Karel and Flink postulate that the volatile components are trapped into microregions of the carbohydrate at sufficiently low water contents. In these microregions the carbohydrate orients during freezing and drying in such a way as to surround the volatile molecules and prevents their escape. The dehydrated solid is amorphous and glassy but highly hydrogen-bonded. This molecular interpretation of volatile "trapping" would correspond to the rapid fall-off of diffusion coefficient found by Thijssen and co-workers for acetone in aqueous solutions of maltodextrin or coffee solids at high solids contents. In fact, the two proposed mechanisms of volatiles loss (selective diffusion and trapping) are quite similar, one being a macroscopic description and the other a molecular description.

Flink and Karel[94] also allowed various volatile components to adsorb onto freeze-dried specimens and then measured rates of desorption in the same evacuated-flask system. The adsorbed volatiles were removed much more readily than was the volatile already present before freeze-drying. This result again is inconsistent with an adsorption theory of volatiles retention. Flink and Karel[94] measured the distribution of remaining volatiles content within freeze-dried samples and found the volatile to be uniformly distributed rather than being concentrated away from the outer surface. This result is also in concord with the Thijssen and Rulkens approach described above since the model of those latter authors would allow for gradients of volatiles content only over distances such as 10 to 100 microns within a given layer between ice crystal columns. In terms of macroscopic concentrations, the Thijssen and Rulkens approach indicates that the volatiles would still be evenly distributed through the dried material.

In a further study of the mechanism of volatiles retention Karel and Flink[A] measured volatiles loss when the freeze-dried specimens were subjected for a prolonged period of time to a controlled relative humidity, higher than the equilibrium relative humidity at the end of drying. For sufficiently low values of this new relative humidity there was no appreciable loss of the remaining volatile organic compound over a period of several days. This was the case for relative humidities below 32% for t-butanol in 10,000 molecular weight dextran

(10% equilibrium water content by weight) and in maltose (6% equilibrium water content). At higher relative humidities the volatile content of the freeze-dried specimen decreased at first and then reached a new, non-zero asymptotic level. Two-thirds of the remaining 2-propanol was lost in this way from freeze-dried maltose held at 61% relative humidity (15% equilibrium water content), and ¾ of the remaining 2-propanol was lost from freeze-dried 10,000-molecular-weight dextran held at 61% relative humidity (19% equilibrium water content). At still higher relative humidities the material lost all of its originally retained volatile without reaching a new, non-zero asymptotic volatiles content.

Karel and Flink[A] interpret the volatiles loss upon humidification to be a reflection of the action of sorbed water entering the carbohydrate matrix and disrupting the hydrogen bonds holding the carbohydrate molecules into the cage structure which had trapped the molecules of the volatile species. They indicate that when the moisture content was sufficient to allow appreciable volatiles loss from a humidified sample there was a visible transition in the freeze-dried cake to a thick, viscous, glassy, amorphous mass. Karel and Flink attribute the occurrence of a new, but lower, asymptotic volatiles content to the persistence of some hydrogen-bonded carbohydrate structure in some regions of the substance, even though the structure has been disrupted enough at other places to give substantial volatiles loss. They found that the tendency for a total loss of volatiles rather than reaching a new non-zero value coincided closely with humidification to a high enough moisture content to give a saturated solution of the sugar in water. This fact suggests that some hydrogen-bonded carbohydrate structure persists up to the point of complete solution. Some humidified samples which had lost some, but not all, of their volatiles contents were subsequently freeze-dried back to a low moisture content and it was found that there was a strong retention of the remaining organic volatile substance during this second freeze-drying step.

Lactose when freeze-dried formed an amorphous solid with good volatiles retention characteristics; however, lactose has the capability to crystallize when rehumidified to a sufficiently high moisture content. In rehumidification experiments involving freeze-dried lactose samples containing 2-propanol as the volatile organic component, Karel and Flink[A] found a slow propanol loss at moisture contents of 8 to 12% before crystallization took place. When recrystallization started, however, there was a more rapid loss of propanol, accompanied by a loss of water caused by the crystallization. Essentially all of the propanol was lost, even though the final water content was low (2 wt per cent). This result points out the need for an amorphous, rather than crystalline, carbohydrate structure for trapping of the volatile component.

The previous experiments were carried out at ambient temperature. Karel and Flink[A] also report the results of humidification experiments made at −18°C, where samples could reach a high moisture content upon humidification and maintain this for a period of time without structure collapse or significant volatiles loss. Eventually the structure would collapse, and at that point in time there would begin to be extensive volatiles loss. As they point out, the volatiles loss would therefore appear to be related to the collapse of the original freeze-dried structure rather than being uniquely related to the moisture content. This sort of behavior and the reaching of a new, non-zero asymptote in some cases are not explained by a simple diffusion model of volatiles loss, in which the diffusion coefficient is taken to be a unique function of water content.

The volatiles retention behavior of a carbohydrate system during freeze-drying thus reflects the entire processing history of the substance and, in particular, reflects the form of molecular aggregation and the degree of crystallization of the solid or concentrated liquid phase formed during freezing and subsequent drying. In this way the results of Karel and Flink are reminiscent of the results of Rey and Bastien[D,6] and MacKenzie[28,29] regarding the effect of the nature of the freezing steps before freeze-drying on the drying rate, the keeping quality of the product and the preservation of structure during freeze-drying. These factors are all dependent upon the actual state of molecular aggregation which is formed in the freezing step and which exists during drying. This state of aggregation can reflect the cumulative effects of all the past processing history and is apparently not necessarily uniquely related to the water content and the temperature. In this respect it is important to note that Karel and Flink[A,94] froze their samples by immersion of a flask in liquid nitrogen, with freezing being essentially complete within 35 or 40 seconds.[95] The volatiles retention behavior which they encountered, therefore, typifies sugar solutions which have been frozen relatively rapidly.

The retention of volatile flavor and aroma compounds during freeze-drying is closely related to the practice of "locking in" volatile flavor substances by dispersing them into a saccharide, which has been carried out for citrus oils and related substances.[96] The factors promoting aroma retention during freeze-drying should be the same as those which make the "locking in" process effective.

Spiess[97] reported taste panel evaluations of aroma for various fruits and vegetables freeze-dried under controlled conditions. Paprika showed maximum aroma retention for the fastest rates of freezing before drying. For various foods the aroma retention was independent of the temperature of the frozen zone during freeze-drying (chamber pressure) until the point of incipient melting was reached. Parsley showed a tendency for poorer aroma retention at higher frozen zone temperatures. Increasing the temperature of the outer, dry surface of the food so as to accelerate freeze-drying caused poorer aroma retention for the substances tested, which included blueberries, raspberries, strawberries, parsley, mushrooms and peas.

Physical and Chemical State of Water in Frozen and Freeze-Dried Foodstuffs

An understanding of the effects of moisture content and processing conditions upon the quality factors characterizing freeze-dried foods requires some knowledge of the ways in which water is bound into foodstuffs. General reviews of this question have been given recently by Rockland,[98] Nemitz,[99] Kuprianoff,[71] Rey[100] and Simatos.[101] Rey[100] identifies four different forms of water binding, as follows:

1. *Free water,* which constitutes the majority of water present in fresh foods and which exerts the full vapor pressure of water at a given temperature.

2. *Water of crystallization,* which is incorporated into solid crystal structures of simple molecules as water of hydration.

3. *Water of constitution,* which is incorporated into macromolecules, such as DNA, in much the same way that water of hydration is incorporated for simple molecules.

4. *Sorbed water,* which is further divided into

   a. *Water of interposition,* located in more or less mobile films, perhaps with a small radius of curvature causing a vapor pressure depression.

   b. *Multilayer adsorbed water,* where water molecules are located near to but not primarily at strong polar adsorption sites.

   c. *Monolayer adsorbed water,* which is primarily adsorbed on a one-to-one basis at strong polar adsorption sites.

The nature of water binding at different water contents in foods can be examined in a number of ways, such as through dielectric constant and loss factor measurements,[99-102] through nuclear magnetic resonance studies[98,100,101] and through calorimetric measurements of the percentage of the total water frozen vs. temperature at different water contents.[102-104]

Probably the most common measurement made to determine the various forms of water present at low moisture contents is the *moisture sorption isotherm,* which is a measure of equilibrium relative humidity (sometimes called water activity) as a function of moisture content. The measurement and interpretation of sorption isotherms for foodstuffs has been recently reviewed by Labuza.[105] Experimental sorption isotherms are available for a large number of food products.[31,87,100,102,106-109] Measurements of *rates* of sorption and desorption in most types of sorption apparatus do not yield primary information about the water binding in the foodstuff since they reflect the heat and mass transfer parameters of the experimental apparatus and the foodstuff.[69]

Moisture sorption measurements when compared to nitrogen sorption measurements[31,87] show the extent to which water molecules penetrate the interior structure of a proteinaceous material. Moisture sorption levels are usually several orders of magnitude higher than nitrogen sorption levels at a given relative humidity of the sorbing gas. This is usually interpreted as indicating that nitrogen covers the open pore surface of the material, while water penetrates inside the solid material to reach free polar groups of individual protein molecules. Thus, nitrogen sorption should be dependent upon the size and shape of the pores left behind by the sublimation of the ice crystals, whereas water sorption should be relatively independent of the pore size. King et al.[31] found that moisture sorption isotherms for freeze-dried turkey meat were effectively independent of the rate of freezing before freeze-drying and, hence, independent of the pore size. Nitrogen sorption isotherms, as reflected by the BET surface area, were highly dependent upon the pore size and initial freezing rate, and the nitrogen surface area correlated well with the pore size distribution measured by mercury intrusion porosimetry. Spiess[87] found that nitrogen surface areas for freeze-dried egg white and potato starch increased with increasing porosity, which was consistent with microscopic measurements of pore sizes. Spiess et al.[60] obtained similar results in comparing nitrogen surface areas to observed pore sizes for freeze-dried maté extract. For water sorption onto freeze-dried beef muscle, MacKenzie and Luyet[110] report an increase in sorbed moisture at any given relative humidity for specimens frozen more rapidly before drying. The change of the amount of water sorbed with respect to freezing rate is still much less than the changes reported elsewhere for nitrogen sorption, however.

Strasser[111] has noted that the amount of water sorbed at a given relative humidity and the size of the sorption-desorption hysteresis loop both tend to decrease with respect to time during storage of freeze-dried beef, and has suggested that the measured sorption isotherm can be used as an indirect quality control monitor.

Deteriorative Chemical and Biochemical Reactions During Drying and Storage

Deteriorative reactions occurring during freeze-drying and subsequent storage of the freeze-dried product have been reviewed by Goldblith and Tannenbaum,[112] Goldblith et al.,[113] Goldblith,[114] Karel[54] and Bimbenet and Guilbot.[115] These reactions range from ones, such as nonenzymatic browning and lipid oxidation, which are relatively well identified and understood, to others which cause loss of color, taste or texture and which are not well identified at all. The effects of the less well known reactions upon product quality are, for the most part, assessed through taste panel tests.

The literature on deteriorative chemical and biochemical reactions is vast and is, for the most part, beyond the scope of this review. In this section primary attention will be given to other reviews and to work specific to freeze-drying which holds the ultimate promise of providing either a quantitative prediction of the effect of freeze-drying conditions and storage conditions upon product quality, or a quantitative constraint upon allowable processing conditions.

Lipid Oxidation

The oxidation of lipids, or fats, is a deteriorative reaction which is particularly prominent for freeze-dried foods because of the very large internal surface areas of freeze-dried materials and because of the very low moisture contents which are usually achieved during freeze-drying.[54,112,113] The reaction is one between oxygen and lipid substances present in the freeze-dried foodstuff. It has been shown for a number of foods[116,117] that the primary lipid contributing to oxidative deterioration is linoleic acid.

Karel[54,113,118,119] measured rates of oxidation of linoleic acid as a function of oxygen partial pressure and found that the surface area exposed to the oxygen is an important variable. Oxidation rates for linoleic acid exposed to oxygen in bulk or in well agitated vessels decreased with decreasing oxygen partial pressure, but when linoleic acid was dispersed on powdered cellulose the rate of oxidation remained high and was independent of oxygen partial pressure down to partial pressures less than 0.005 to 0.01 atm. The lipid distribution in freeze-dried foods is expected to approach this colloidally dispersed case, especially if there has been melting of the lipids during drying.[115] These results show the need for very complete removal of oxygen before packaging of many foodstuffs and for the use of relatively oxygen-impermeable packaging materials.

Goldblith et al.[113] point out that the slow rates of desorption of oxygen to be expected from freeze-dried foods[69] mean that a simple evacuation-plus-nitrogen-flushing technique carried out before packaging may in actuality leave behind an oxygen content of 5 to 15% within the package gas space once desorption equilibrium has been approached. Goldblith et al.[54,113] also point out that the degree of oxygen impermeability required of a packaging material in order to minimize deterioration through lipid oxidation over long periods of time is such as to preclude polymer films for packaging in many cases. Spiess et al.[60] suggest that measured nitrogen surface areas of freeze-dried materials can serve as an index of the susceptibility to gas damage from oxygen during storage, with lower nitrogen surface areas leading to lesser oxidation tendencies.

Various oxygen scavengers for inclusion with freeze-dried foods in the package have been suggested and evaluated.[113,120] The classical system of glucose and enzyme glucose-oxidase is unsatisfactory because the water content required is higher than that desired within the package.[113] A system of hydrogen and a palladium catalyst has been used with some success,[120] but the expense and complexity of such an approach are major problems. Exclusion of light is helpful for reducing oxidation rates but is not adequate in itself.[113]

Labuza and co-workers[117,121,122] have examined the kinetics of lipid oxidation reactions and the dependency upon moisture content in model systems and in freeze-dried salmon. It has been known for some time that lipid oxidation is favored by lower moisture contents; this is one of the reasons for the importance of the reaction for freeze-dried foods. Labuza and co-workers conclude that the effect of water in inhibiting lipid oxidation comes from two sources, (1) hydration and thereby deactiva-

tion of metals, notably cobalt, which catalyze the oxidation reaction, and (2) formation of hydrogen bonds with hydroperoxides, thereby preventing their entering into initiation reactions. Maloney et al.[121] and Martinez and Labuza[117] analyzed the kinetics of the lipid oxidation reaction through a reaction model in which every oxygen molecule reacts with a linoleic acid molecule to form a molecule of linoleic acid hydroperoxide, with the linoleic acid being present in sufficient excess so that its concentration is effectively constant.

Non-Enzymatic Browning

Non-enzymatic browning reactions, or Maillard reactions, are deteriorative reactions which involve the carbonyl group of a reducing sugar and the amino group of an amino acid or protein. The reaction products are brown, polymeric, insoluble substances. The deterioration occurs through color change of the product, the formation of off-flavors and, in some cases, the loss of nutritional value since lysine and ascorbic acid, for example, participate in these reactions. Quantitative rate data are valuable since Maillard reactions tend to occur most rapidly at intermediate water contents in foods; and, hence, the extent to which these reactions occur is a variable function of the drying conditions.

The chemistry of non-enzymatic browning reactions has been reviewed by Lea,[123] Hodge,[124] Ellis[125] and Reynolds.[126] More recently, Song, Chichester and Stadtman[127] have reported a very extensive, quantitative kinetic analysis of the browning reaction between d-glucose and glycine in bulk solution. They are able to account for the observed induction period of the reaction through the proposed mechanism and kinetics. There are differences in rate to be expected between bulk solution and moist solids. For example, Loncin et al.[106] found the peak browning rate between alanine and glucose in glycol-water solution to occur at very low water mole fractions; however, for milk powder they found a peak browning rate at a moisture content corresponding to equilibrium with a 40% relative humidity.

Kluge and Heiss[B] report an interesting effort to relate the amount of chemical degradation quantitatively to conditions of temperature and moisture content during drying. They measured rates of non-enzymatic browning for samples of a model solution containing glucose and glycine in a molar ratio of 1:2 deposited upon cellulose powder with different moisture contents. The progress of the browning reaction at fixed temperature and humidity was followed by means of UV spectrophotometry. They noted that the reaction rate appeared to become greater as time went on, indicating an autocatalytic nature of the reaction. An induction period for the reaction has also been reported by others.[127]

Kluge and Heiss present results of the kinetic measurements at different temperatures and relative humidities in terms of a quantity called t_{zul} (Zeit zulässige = allowable time), which is the time necessary to achieve a specified optical permeability under conditions of constant temperature and relative humidity. This optical permeability corresponded to a specific extinction of 0.4, which marked the beginning of an appearance of a yellow color in the substance. t_{zul} would be inversely proportional to the reaction rate constant if a simple reaction of uniform order occurred across the range of concentrations encountered.

Kluge and Heiss[B] noted that the peak reaction velocity occurs at about 2% water content. This value of moisture content for the peak rate is less than most of the other peak-rate moisture contents which have been noted for foodstuffs.[106,115,123] Such behavior may reflect a vagary of the particular supported system which they studied, or it may come from their carrying the reaction further than would correspond to the amount of deterioration occurring during most drying operations. The specific extinction of 0.4 selected by Kluge and Heiss for t_{zul} appears, by their Figure 4, to correspond to a disappearance of about 7% of the glucose present.

Kluge and Heiss[B] devised an approach for predicting the cumulative amount of browning occurring during drying under changing conditions of temperature and relative humidity. The approach is similar to that used for analyzing sterilization under changing temperatures. From their derivation they conclude that the maximum allowable extent of browning, corresponding to a specific extinction of 0.4 in

their case, will be reached at a time, t_e, after the start of drying given by

$$\int_0^{t_e} \frac{dt}{\left[t_{zul}\right]_{X,T}} = 1 . \qquad (22)$$

In Equation 22 $[t_{zul}]_{X,T}$ will vary with time and will be a function of the particular moisture content, X, and temperature, T, prevailing at any spot in the material being dried at any time. $[t_{zul}]_{X,T}$ is the time required to reach the limiting amount of browning at constant values of temperature and moisture content equal to the X and T at which t_{zul} is evaluated.

Kluge and Heiss derive Equation 22 through an analysis based upon first-order reaction kinetics in which the optical permeability is directly proportional to the amount of the remaining reactant in which the reaction is first order. Since the optical permeability decreases in proportion to the amount of colored products formed, such a kinetic mechanism is quite unlikely.

Kluge and Heiss state that their analysis would apply as well to reactions of other orders; however, it may readily be shown for the general case that Equation 22 using values of t_{zul} observed at constant X and T will only be valid if the kinetic rate expression may be put in the form:

$$\frac{dM}{dt} = -k(X,T) \cdot f(M) , \qquad (23)$$

where
    M = a direct measurement of the amount of reactant remaining (i.e., the optical permeability)
    t = time
    X = moisture content
    T = temperature.

k and f are arbitrary functions of X and T and of M, respectively. The only assumptions made with regard to reaction mechanism or reaction order are that the functions used apply to the entire range of temperature and moisture content and that the two functions k and f are separable. By "separable" it is meant that the kinetic expression may be written so that k is independent of M and so that f is independent of X and T.

Unfortunately, this condition of k(X,T) and f(M) being separable does not seem to be met by the browning reactions. One would expect that for low conversions (i.e., low loss of reactants) it might be allowable to use such a rate expression with f(M) held constant at a value corresponding to the initial reactant concentrations. For browning reactions in solution, however, Song, Chichester and Stadtman[127] have shown that it is necessary to incorporate concentration levels of reaction intermediates and products into the rate expression in order to explain observed rate data. Their rate expression[127] takes the form

$$\frac{dB}{dt} = [k_2\{k_i(G_o)/k_2'(I)\}^{\frac{1}{2}}$$

$$+ k_3(g_o - B)] \cdot [I - B] , \qquad (24)$$

where

G_o = concentration of sugar (glucose)
g_o = concentration of amino acid (glycine)
I = concentration of reaction intermediates
B = concentration of brown products

The various k's are reaction rate constants and are exponential functions of temperature. Only if all the k's have the same activation energies would the separable-function criterion of Equation 23 be met. Similarly complex kinetics would be expected for the reaction in a supported medium, as in drying. For a supported medium, the k's would have to be identical functions of moisture content as well as temperature in order for the separable-function criterion to be met. The inclusion of the intermediates and products in the mechanism is necessary to account for the induction period which, in turn, is important in any analysis of browning during drying. Thus it appears that the procedure put forward by Kluge and Heiss for determining drying conditions leading to a maximum allowable amount of browning is invalidated by the underlying assumption of separable functions in the kinetic expression.

As was acknowledged by Kluge and Heiss,[B] a further complication arises from the fact that the histories of temperature and moisture content will be different at different points in a substance being freeze-dried. Thus an analysis of the extent of browning during drying should really be based upon local, rather than average, conditions of temperature and moisture content. If a kinetic expression, similar to Equation 24 but for browning in supported media, were known, one could then use computer solution of the kinetic differential equation to analyze the extent of browning to be expected at different points within a substance for various conditions of freeze-drying.

Despite the limitations of the specific analysis offered by Kluge and Heiss, two useful conclusions for freeze-drying may be derived qualitatively from a consideration of the interaction of browning kinetics with drying conditions:

1. From their results Kluge and Heiss derive Arrhenius activation energies for the browning reaction lying between 27 and 32 kcal/gmole, which are similar to apparent activation energies reported by others for the browning reaction,[115,127] The activation energies for other degradative reactions appear to be of a similar order of magnitude, whereas the temperature dependence of rates of water removal during drying typically corresponds to apparent activation energies of 10 kcal/gmole or less.[69] Hence, the degradative reactions vary more sharply in rate with changing temperature than does the drying rate. As a result the amount of degradation for a given amount of water removal should be least for drying carried out at the lowest temperatures. This is another incentive for low-temperature drying, and particularly for freeze-drying, as compared to other types of drying. This incentive has also been noted by Thijssen and Rulkens.[59]

2. In addition to the low temperature of operation, one of the great benefits of freeze-drying for avoiding browning in those substances which show a peak browning rate at intermediate moisture contents will be the relatively rapid transition from a fully hydrated condition to a low moisture condition as the ice front passes any particular point within the substance being dried.

Other Reactions

The occurrence of enzymatic reactions in dried foods during drying and storage has been reviewed by Acker.[128,129] Enzymatic reactions can cause food degradation in a number of ways, including oxidative enzymatic browning of cut fruits, cleavage of starch molecules with the production of simple sugars, hydrolysis of pectins and hydrolysis of lipids.[115] Enzymatic degradation activity in dried foods and during drying processes proceeds more rapidly at higher water contents,[115,129] and one of the benefits of food dehydration is the suppression of enzymatic activity while the food is in the dry state. A significant amount of enzymatic reaction can occur in the frozen state, although the activity in the unfrozen state is usually greater. Studies of enzymatic reactions in the frozen state have been reviewed by Lund et al.[130]

Freeze-drying usually gives better performance than other drying methods in the preservation of vitamins and other nutritive values present in foodstuffs. A good review of studies of the loss of vitamins and nutritive values during freeze-drying is given by Goldblith and Tannenbaum.[112] Additional work on vitamin loss during freeze-drying, prior processing and subsequent storage is presented by Spiess.[97]

Color changes can come from a number of reactions in addition to the obvious color changes from enzymatic or non-enzymatic browning reactions. Color changes are generally undesirable, even if they are cured upon rehydration, because they reduce the visual attractiveness of the freeze-dried products. Color changes can also signify other changes; for example, the loss of β-carotene produces a color loss in both carrots and paprika in addition to being a loss of a potential vitamin. Lusk, Karel and Goldblith[54,114,131] have measured rates of loss of astacene pigment in freeze-dried shrimp and salmon during storage. Bengtsson and Bengtsson[120,132] have found differences between cooked and raw beef with regard to the effect of processing variables during freeze-drying on product color and appearance. Spiess[97] measured color changes caused by freeze-drying in strawberries, carrots and parsley, using a tri-chromatic color analyzer. For freeze-drying of coffee it has been found that the product color is closely related to the freezing conditions before drying, as was indicated earlier.

Protein denaturation refers to a loss in water solubility and a change in intermolecular binding of proteins which can be caused by factors such as high temperatures and high salt contents. Protein denaturation is a problem during drying processes and causes poor rehydration and changes in texture. Denaturation is a complex phenomenon[134] and will be considered here only from the standpoint of the maximum temperature limitation which it imposes on the outer, dry surface. Denaturation threshold temperatures lie in the range of 40 to 80°C, depending upon the protein and the pH.[115,134] In a study of freeze-drying of ox muscle MacKenzie and Luyet[133] found that denaturation came from high surface temperatures (80°C in their case) rather than from the fact of water removal itself. Nemitz[102] found that the temperature threshold for the denaturation of the albumin in egg white was 60°C, as long as the equilibrium relative humidity was above 20%, but that the threshold temperature rose linearly toward 125°C at equilibrium relative humidities below 20%. Aitken[135] found that surface temperatures below 60°C during freeze-drying did not alter the appearance or organoleptic properties of pork. A quality degradation was noticed by a taste panel for a surface temperature of 80°C during freeze-drying, and a decrease in protein solubility and glucose content (Maillard reaction) could be detected analytically at surface temperatures between 60° and 80°C.[135]

In fish and meats texture and tenderness are related to changes in the underlying protein structure and/or incomplete rehydration. Tenderness of rehydrated freeze-dried meat has been examined by Miller and May[136] and Tuomy et al.[137] among others.

Shrinkage

One of the foremost advantages of freeze-drying as a dehydration method for particulate foodstuffs is the effect of the ice structure in minimizing shrinkage of the product. The reduced shrinkage results in a porosity which facilitates vapor escape and which enables rapid and nearly complete rehydration.

Kluge and Heiss[B] report the results of extensive measurements of shrinkage of lean beef and of potato dice during freeze-drying under vacuum. The heating-plate temperature during a run was adjusted so as to hold a constant ($\pm$ 2°C) measured temperature of the frozen zone during drying, and different frozen-zone temperatures were maintained in different dry-

FIGURE 12

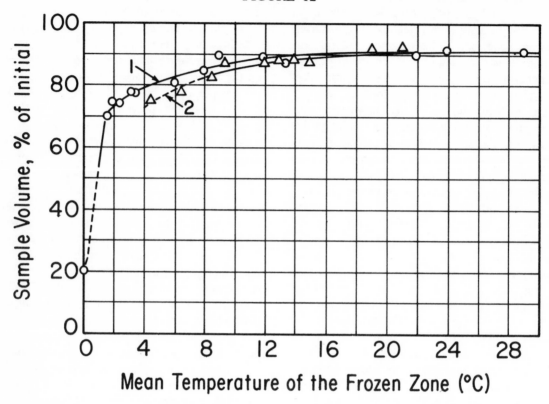

Shrinkage of potato dice after freeze-drying at different frozen-zone temperatures. Freezing rate before drying = 2.6 cm/hr (Curve 1) = 1.2 cm/hr (Curve 2.) After Kluge and Heiss.[B]

ing runs. The results for potato dice are shown in Figure 12 and are presented as the ratio of the piece volume after drying to the volume before freezing and drying. Similar results were obtained for lean beef. Kluge and Heiss[B] point out the similarity between the shape of the shrinkage curves shown in Figure 12 of this paper and the corresponding curve for lean beef, on the one hand, and the shape of the curves of per cent frozen water vs. temperature found by Riedel,[104] on the other hand. The shrinkage appears to be very nearly proportional to the percentage of the water which is unfrozen at the temperature of the frozen zone. The indication is that the frozen water provides the structure preventing collapse upon drying, but that portions of the substance occupied by the unfrozen water do collapse when that water is removed. Kluge and Heiss indicate that the shrinkage occurs during the sublimation portion of drying and not to any great extent during terminal drying

to remove the last moisture. Shrinkage of potato pieces frozen at two different rates is shown in Figure 12, and the shrinkage is greater for a slower freezing rate before freeze-drying. Kluge and Heiss suggest that this result comes from more cell shrinkage and structural damage during freezing because of the formation of larger intercellular ice crystals.

King, Lam and Sandall[31] reported measurements of shrinkage after freeze-drying and after rehydration for several pieces of turkey breast meat. Referred to the frozen volume before drying, the shrinkage after drying ranged from 13 to 22% and the shrinkage after rehydration ranged from 9 to 14%. These figures would be reduced by about 7% (i.e., 6 to 15% and 2 to 7%) if the shrinkage was referred to the volume before freezing. From their data King et al.[31] concluded that shrinkage was promoted by higher temperatures of the outer piece surface and higher frozen-zone temperatures during drying. Since in the work of Kluge and Heiss, higher frozen zone temperatures were obtained through higher outer surface temperatures, the results of King et al.[31] are in agreement with their conclusions regarding shrinkage.

Beke[23] measured the expansion of pork slabs upon freezing and the amount of shrinkage caused by subsequent freeze-drying. The expansion upon freezing ranged from 8 to 8.5% and the shrinkage caused by drying as compared to the frozen state ranged from 15 to 23%, or 7 to 15% as referred to the initial volume before freezing. These amounts of shrinkage agree well with the results for beef and turkey; however, Beke found the shrinkage to be least for a 50°C outer piece surface temperature as compared to higher and lower temperatures, a result which is at variance with the behavior reported by Kluge and Heiss and King et al.

Rehydration Ratio

The rehydration ratio is a simple test for physical damage to a foodstuff during freeze-drying. The test generally carried out is to soak the freeze-dried product in water at ambient temperature for a fixed length of time, to blot it and then to weigh it. The rehydration ratio is the ratio of the weight of water regain to the weight of water lost during freeze-drying. Poor rehydration reflects sample distortion and denaturation of proteins. Again, rehydration is a field in which much work has been done,[138] and the purpose here will be only to relate reported studies of the dependence of rehydration upon processing conditions.

Beke[23] for pork and King, Lam and Sandall[31] for turkey found that conditions leading to a lack of shrinkage during freeze-drying also led to an improved rehydration ratio. King et al. found rehydration ratios of 0.87 to 0.95 for freeze-dried turkey slabs of ¼ and ⅜ inch thickness when immersed in water at 25°C for 20 minutes. Karel[54,114] found that the rehydration ratio of freeze-dried haddock fillets decreased for increasing outer surface temperature (and, hence, frozen-zone temperature) during freeze-drying. Conversely, Karel[54] reports better rehydration for salmon freeze-dried at higher chamber water vapor pressures and, hence, higher frozen-zone tem-

peratures. Goldblith et al.[113] report extensive studies of rehydration for freeze-dried shrimp. A higher surface temperature and resultant higher frozen-zone temperature during freeze-drying gave reduced rehydration. Increased cooking time at 180°F or 212°F before freeze-drying also gave poorer rehydration for shrimp, especially for cooking at 212°F. Increased temperature of the rehydrating water gave poorer rehydration for both shrimp and salmon, over the range 40 to 212°F, presumably because of protein denaturation by the rehydration water. Studies of the amount of rehydration vs. time showed that most of the rehydration water is picked up within the first five minutes after immersion.

Drip- or Water-Holding Capacity

The water which reenters a freeze-dried food is more loosely held than the water which was in the original food. A substantial amount of water can be forced back out by simple expression or even by gravity, much as water is lost from a sponge. This problem would seem to result from the fact that rehydration water enters the pores left by the ice crystals, but does not reenter the microstructure of the foodstuff to as great an extent as was the case before freezing. One would therefore expect that the tendency for loss of rehydration water through expression would be greatest for specimens that were slow-frozen before freeze-drying, since slow freezing forms the largest ice crystals and, therefore, leaves the largest pores after freeze-drying. Karel[54,114] found that the water-holding capacity (defined as retained rehydration water after being subjected to a pressure of 100 psi between sheets of filter paper) for haddock fillets was less for higher surface temperatures and higher frozen-zone temperatures during freeze-drying.

Optimal Residual Water Content

The optimal water content to be left in a freeze-dried food reflects a compromise among the various degradative phenomena which may occur during freeze-drying and during storage and will be different for different foodstuffs. Salwin[107] recommended that the equivalent monomolecular layer moisture content, as determined from a BET analysis of the moisture sorption isotherm, would be a desirable final moisture level in most cases. This corresponds to between 2 and 5% residual moisture for most food products. On the other hand, Martinez and Labuza[117] found that an equilibrium relative humidity of 32%, or about 7% residual moisture (well above the BET monolayer value), gave the best product stability for freeze-dried salmon because of lipid oxidation and astacene pigment loss during storage at lower water contents. Rockland[98] has analyzed degradation data for various dehydrated food products and suggests that maximum storage stability will occur for residual moisture contents in the equivalent BET multilayer sorption range of the isotherm, specifically in the intermediate, flat portion of the isotherm where the change in the equilibrium moisture content with changes in relative humidity is least. This corresponds to higher moisture contents than are usually achieved by freeze-dryers and can obviate the need for a lengthy terminal drying period.

PROCESSING APPROACHES

Market and Production Status

The growth of the freeze-drying market at present is rapid, largely because of the success of freeze-dried instant coffee. There appears to be no comprehensive recently published review of the market and plant-size status of freeze-drying in the United States; however, the status in Europe has been recently covered by Spicer[139] and by Bengtsson.[140,141] Earlier, the status of freeze-drying in the United States was analyzed by Tease,[142] by Smith,[143] and in various reports from the Economic Research Service of the United States Department of Agriculture.[144,145]

Processing Problems

Referring back to Figure 4 and to the previous discussion, we may see that the processing problems which must be overcome in any freeze-dryer design fall into several categories:

1. Supply of an amount of heat sufficient to cause sublimation of the water within the foodstuff.

2. Efficient transfer of this heat from the heat source to the surface of the foodstuff.

3. Transfer of the heat through the foodstuff to the sublimation front, either through the dry layer or through the frozen layer, or through both.

4. Transfer of the generated water vapor from the sublimation front to the outer surface of the foodstuff, through the dry layer.

5. Efficient transfer of the water vapor from the surface of the foodstuff to the moisture sink.

6. Removal of the water vapor at the moisture sink, along with some kind of regeneration or renewal of the moisture sink.

7. Maintenance of a low enough temperature of the frozen zone so as to minimize structural collapse and deteriorative reactions during drying.

8. Maintenance of a low enough temperature of the outer, dry surface of the foodstuff so as to minimize deteriorative reactions at that point during drying.

9. Maintenance of a low enough pressure level so as to preclude excessive rate-limits imposed by molecular diffusion of water in the gas phase at any point.

10. Effective retention of volatile flavor and aroma species where this is important, by means such as low temperature, rapid formation of a selective diffusion zone, etc.

11. Facilities for introduction and removal of the material to be dried with a minimum of complexity and labor time.

Satisfying these various requirements requires compromising a number of factors against one another. Also, each of these requirements can be satisfied in a number of different ways, making the number of possible processing approaches quite large. Obviously, some combinations of ways of satisfying the various needs are much more attractive than others, and it is the goal of the freeze-drying process designer to find the particular combination which is best. Kröll[146] has classified and analyzed dryers in general according to their means of satisfying different processing requirements, and Clark[24] has considered this combination problem for freeze-drying of foodstuffs, in particular, with emphasis on those combinations of ways of fulfilling the different requirements which interact favorably with one another.

Processing approaches which have been used in practice or are known to be in the development stage have been summarized in a collection of papers edited by Cotson and Smith,[8] by Miner,[147] and, more recently, in a pair of reports available from the Svenska Institutet for Konserveringsforskning (SIK).[140,141]

Conventional Plants

Conventional freeze-drying plants in use today generally support the food material to be dried in particulate form on a series of trays within the drying chamber. Such a dryer is shown schematically in Figure 13. Heat is supplied by a circulating heat transfer agent within platens which may be in direct contact with the trays or may transmit heat by radiation. Steam may also be used as a heating agent in radiant-heat freeze-dryers. Although steam jet ejectors are used to some extent in Europe for removing water vapor as well as inerts, water vapor removal is generally accomplished by ice formation onto chilled, metal surface condensers which are defrosted after

the conclusion of freeze-drying. The condenser unit is backed up by a vacuum pump or steam jet for inerts removal, and may be located within the drying chamber[148] or in one or more separate chambers which may be closed off intermittently for defrosting.[149,151] Extensive and detailed descriptions of such plants and the options available within them are available in the literature.[8,140,141,147,148,150]

Continuous Processes

Nearly all large-scale freeze-drying operations to date have been batch-operated as a result of the problems of feeding and removing materials from a vacuum system, of loading trays evenly, etc. As in any large-scale processing operation, there exists considerable incentive for development of a truly continuous, large-scale freeze-drying apparatus if such a device can be made to operate reliably and economically. A particular incentive comes from the prospect of balancing the load imposed on the water vapor condensation system and the vacuum system. In a batch process the water vapor evolution rate from the foodstuff is quite high at the start of drying and becomes less as drying proceeds. The condenser system must be designed to handle the

FIGURE 13

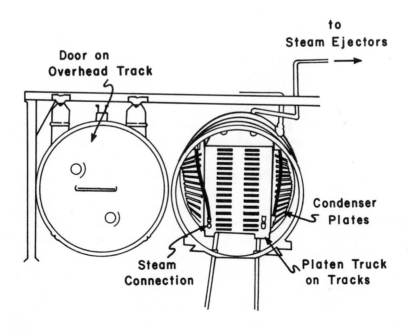

Typical commercial freeze-dryer.

maximum water vapor removal requirement.

The first approach toward a continuous freeze-drying process in large-scale practice has been in the Leybold System,[151,152] in which an overhead rail conducts layered sets of trays through a long tunnel. Vacuum locks, intermittently operated, are located at the inlet and the outlet of the tunnel to allow for supplying the food to be dried and for removal of the product. Because of this feature the process might best be called semicontinuous. Another approach toward continuous freeze-drying is the process devised by Rockwell et al.[153,154] and demonstrated on a pilot scale, in which particulate foods tumble along a slowly rotating, inclined tube, with the tube being polygonal in cross-section ꓳ as to provide good tumbling and mixing action. Heat is supplied continuously from a steam chamber through which a number of such tubes pass, and water vapor is removed by a conventional vacuum system. In another variant of this process[155] the heat is supplied through finned tubes which act as baffles to promote mixing of the particles of the substance being dried within a rotating chamber.

Another approach to continuous freeze-drying, described by Pfluger and Elerath,[156] uses gravity loading and unloading through vapor locks into many vertical cells in parallel which provide means of heating distributed vertically along each cell and which provide small dimensions across which vapor can escape from the material being dried. This approach is designed to overcome the problems of continuous loading and of distribution of material for even drying within a continuous freeze-dryer. Stinchfield[157] describes a continuous freeze-drying process for liquids involving freezing and drying on a moving belt, and Fuentevilla[158] presents a continuous process utilizing vibration to provide particle motion along the drying path.

Water Vapor Removal and Vacuum Systems

The criticality of the water vapor removal system in a freeze-drying process may be demonstrated by the statement that, typically, between 60 and 95% of the weight of the food put into a freeze-dryer is water, which will expand by a factor of perhaps 10^7 in volume upon vacuum sublimation. Very large volumes of water vapor must be handled and removed. In a tray type freeze-dryer using a chilled metal surface condenser the condenser surface area must typically be of the same order of magnitude as the tray area. The tray area is designed for shallow and even loading of the foodstuff upon it, and it is imperative that as much attention be given to achieving a smooth and even loading of desublimed ice upon the condenser surface.

Methods which have been used or tested for vacuum generation and water vapor removal are summarized and described by Rowe.[159,160] These include multistage steam jet ejectors, various sorts of vacuum pumps, refrigerated metal surface condensers, liquid and solid desiccants, and cold, ice-immiscible liquids. Barrett, Laxon, and Webster[161] have compared the economics and design of steam jet ejector systems and refrigerated condenser (plus vacuum pump) systems in various freeze-drying situations. Refrigerated condensers are currently used in most large-scale applications.

In the design of a refrigerated metallic-surface condenser system, it is important to provide for a minimum pressure drop of water vapor between the food surface and the condensing surface. This calls for large vapor transfer lines from the drying chamber to the condenser chamber, or for putting the condenser surfaces inside the drying chamber. If the condenser is inside the drying chamber the design should guard against excessive heat loss by radiation from the heating surfaces direct to the condenser surface. It is also important to design the condenser system so that the condenser surface will frost up more or less evenly. A system of baffles, forcing all vapors to flow over the condensing surfaces on their way to the pump, has been found to improve operation and provide more efficient utilization of the entire condensing surface.[162] Another problem with refrigerated metal surface condensers is that they must be shut down periodically for de-icing. Togashi and Mercer[163] have proposed the rapid application of heat for fast de-icing, and Tyson[164] has put forward a system in which rapid de-icing is accomplished by pouring hot water or glycol over the condenser surface.

Liquid desiccants, or absorbents, such as lithium chloride brines were investigated by Tucker and Sherwood[165] for vacuum dehydration processes. Water vapor removal with such absorbents would be continuous, but drawbacks of the process are potentially slow absorption rates limited by liquid phase mass transfer resistance, the need for supplying the full latent heat of desorption for all the water removed in order to regenerate to the absorbent, and the effect of the heat of absorption in raising the absorbent temperature. The rise of absorbent temperature necessitates very high rates of absorbent recirculation or else requires that the absorbent be passed in thin films down a refrigerated metallic surface. The absorbent reduces the equilibrium partial pressure of water vapor and thus makes the temperature of refrigeration required less severe, but refrigeration of the absorbent is still required. Rowe[166] has investigated the use of wiped-surface falling films of glycerol and ethylene glycol for water vapor removal in freeze-drying in an effort to accelerate the liquid-phase mass transfer coefficient.

In an effort to circumvent the need for supplying heat to regenerate liquid absorbents, there has been some effort to use as a condensing medium a cold liquid which is immiscible with ice. Thuse[167] and Eolkin[168] describe systems wherein water vapor is condensed onto a spray or shower of a cold, immiscible liquid. There are certain restrictions upon a liquid to be used for this purpose; for example, it must have a very low vapor pressure without having a high viscosity, and either the liquid must be compatible with food or else the condenser chamber must be reliably isolated from the drying chamber. The liquid must also be cheap and/or highly insoluble in water so that any liquid loss with the ice will not destroy the economics of the process. Specific liquids mentioned by Thuse[167] are tetra-2-ethyl-hexyl silicate, di-2-ethylbutyl adipate and di-2-ethylbutyl azelate, whereas Eolkin[168] indicates only that an "oil" would be used. An immiscible liquid spray system requires a very large circulation rate because the temperature rise of the droplets due to the heat of condensation limits the ice loading to 1 or 2%, at most. Kumar[169] has investigated the use of a very nonvolatile but nonviscous immiscible liquid flowing down a refrigerated surface as a condensing medium. Since the liquid in this case serves solely to transmit heat from the condensed ice to the refrigerated metal surface while removing the ice continuously, the liquid circulation rate can be much less than for a spray. Suitable liquids were found to be a low-molecular-weight silicone oil, certain Freons and alkylated aromatic hydrocarbons.

The use of solid desiccants has also been investigated. Rowe[166] suggested using molecular sieve particles to absorb the peak load during batch freeze-drying, thereby reducing the size of the main condenser unit which would be required. Molecular sieve is the preferred adsorbent since a high water capacity at a very low equilibrium partial pressure of water vapor is required, and molecular sieve has an adsorption isotherm shape which fulfills this criterion better than is the case for other solid desiccants.[25,170] Saravacos[170] used molecular sieve to take up the water vapor generated during the freeze-drying of apple slices ½ inch thick. The time required for freeze-drying was about 12 hours. Saravacos noted that, in principle, the heat of sorption of the water vapor onto the molecular sieve could equal the heat requirement for sublimation. Strasser[171] placed frozen food, wrapped in cheese cloth, into a bed of silica gel desiccant and evacuated the chamber. Freeze-drying would then occur if the walls of the vessel were kept at a temperature of 0 to 10°C, and most of the water in the foodstuff was transferred to the solid desiccant. Earlier, Graham et al.[172] had developed a process for the freeze-drying of biological substances in which particulate frozen material (vegetative bacteria in their experiments) was tumbled with silica gel particles in an evacuated cannister.

Conductive Heat Supply

When the heat of sublimation is provided by radiation to the outer, dry layer of a food being freeze-dried under vacuum conditions, the process is almost always rate-limited by heat transfer, either by the external heat transfer coefficient for the radiation or by conduction through the dry layer of the food. In addition, radiant heating is uneven to some extent, depending upon the view factors and temperatures seen by different food pieces. As a result some pieces will be overdried before other pieces are adequately dry.

One approach to overcoming the variability of radiant heating and the rate limitation of heat conduction through the dry layer is to provide the heat by conduction through a frozen zone of the food to the sublimation front. This procedure requires good thermal contact between the foodstuff and a contact heating plate, which is usually the bottom of a tray. The surface area through which vapor can escape is less and it is necessary to hold the temperature of the heating plate below the melting point of the material being dried. Even so, the overall time required for drying is generally less than can be achieved by supplying heat through the dry layer, except possibly for the need for removing a greater amount of sorbed water from the "dry" layer during the terminal stages of drying. The potential benefit of supplying heat by conduction from a contact plate through the frozen layer was recognized by Harper and Tappel[9] and by many others since. The drawback to this method is the necessity for maintaining good contact between the food and the heating plate since, if a gap develops, there will be vapor escape and formation of an insulating dry layer adjacent to the heating plate. Because of the low plate temperature held to prevent melting, drying adjacent to the heating surface will result in a much lower rate of heat supply than would be achieved by supplying heat through the outer dry layer. Good contact with the heater can be obtained with frozen liquid foodstuffs, but even these are usually dried commercially from the ground, particulate state because of the greater particle surface area obtained.

The AFD (Accelerated Freeze-Drying) process, developed by the British Ministry of Agriculture, Fisheries and Food at Aberdeen, used conductive, contact heating to supply the heat for freeze-drying. As reported by Hanson[173] and by Cotson and Smith,[8] this process supplied heat through the dry layer from expanded metal inserts, which provided contact at some points for heat conduction and, at the same time, provided an open path

for easy vapor escape. The expanded metal inserts fit between the food and the heater plates. A pressure of approximately 4 psi was imposed upon the heating plates to promote good thermal contact with the food. A similar process has been described by Oldencamp and Small,[174] who propose supplying heat by conduction from inflatable platens made of a flexible material, such as rubber. A heating fluid, such as ethylene glycol, is supplied to the interior of the platens and to heating plates. Again, expanded metal inserts could be used to provide a path for vapor flow.

Yet another scheme which has been put forth for supplying heat conductively comes from Smithies and co-workers,[175,176] and involves penetrating heated spikes into the material undergoing freeze-drying.

Freeze-Drying at Higher Pressures

The need for maintaining a vacuum in freeze-drying processes carries with it the need for vacuum-tight chamber construction, vacuum-tight doors for loading and unloading, and vacuum generating equipment. Also, at the levels of vacuum ordinarily used in freeze-drying it is not possible to supply heat by convection. As a result heat must be supplied to the food pieces by radiation or conduction, both of which are more complicated than the convective heating processes often used for ordinary drying at higher temperatures.

Atmospheric Pressure

Freeze-drying is still thermodynamically possible at higher pressures. The only absolute requirement is that the partial pressure of water vapor in the drying chamber be kept low enough to provide a mass transfer driving force for removal of water vapor which is generated at a partial pressure in equilibrium with the frozen material. If an inert gas is present, there is no limit to the absolute pressure level. Freeze-drying at atmospheric pressure has been accomplished for years by those who hang their wash out on the line on a cold winter day. Meryman[177] demonstrated freeze-drying at atmospheric pressure in the laboratory and suggested[178] that a process could be built around convective freeze-drying in a cold, circulating air stream, with the water being removed by molecular sieve desiccant or by a refrigerated condenser. Lewin and Mateles[179] successfully freeze-dried carrot and chicken pieces in a circulating loop of cold gas at atmospheric pressure, with water removal by silica gel. Freeze-drying at atmospheric pressure has also been investigated by Woodward[180] and by Sinnamon et al.[181]

The great drawback to freeze-drying at atmospheric pressure is its very slow rate. For example, Lewin and Mateles[179] found that 30 to 40 hours were required to adequately dry discs and cubes of carrot 8 mm thick. The very slow drying rate at atmospheric pressure results from the bulk gas diffusivity being inversely proportional to the total pressure. By Equation 20, such a high pressure gives a very low value of D', which by Equation 9 gives a very low value of k_{gi}. Since the partial pressure driving force, $p_{fw} - p_{ew}$, in Equation 4 is limited to 2 or 3 mm Hg at most, no matter what the total pressure, the very low value of k_{gi} gives a very low drying rate. Hence, some degree of vacuum is necessary in order to provide a commercially acceptable drying rate.

Increased Pressure of Water Vapor

From Figure 2 it should be recalled that the thermal conductivity of freeze-dried foodstuffs increases with increasing pressure, starting at about 0.1 mm Hg and leveling off again at about 500 mm Hg. As long as the heat is supplied through the dry layer and the drying rate is internal heat transfer-limited, an increase in thermal conductivity can cause the drying rate to increase. Ehlers, Oetjen, Hackenberg, Moll and Neumann[182,183] developed a process in which a throttling valve is placed between the drying chamber and the condenser so as to maintain an appreciable pressure (up to 1 or, at most, 2 mm Hg) of water vapor in the drying chamber and thereby increase the thermal conductivity in the dried layer. The drying rate was reported to be somewhat improved as a result.

Optimum Pressure

Sandall, King and Wilke[E] analyzed the interaction of the heat and mass transfer processes within turkey meat during freeze-drying, based upon the theoretical analysis presented previously and based upon their experimental data for drying rates, thermal conductivity and effective diffusivity. Relative maximum attainable rates of freeze-drying as a function of chamber pressure are shown in Figure 14 for freeze-drying in the direction of the fiber and in the direction perpendicular to the fiber. The rates are referred to the rate at zero pressure parallel to the fiber and are based upon the assumption of negligible external mass transfer resistance. The breakpoint on each curve corresponds to the transition from internal heat transfer control to internal mass transfer control. At low pressures the process is internal heat transfer controlled, and the maximum allowable surface temperature of 60°C is the rate-limiting factor. The curves between zero pressure and the breakpoint represent the compensating effects of two changing factors—the thermal conductivity, which increases with increasing pressure, and the thermal driving force, which decreases with increasing pressure as the temperature of the frozen zone rises. At pressures above the breakpoint the process is internal mass transfer controlled, and the drying rate is inversely proportional to the total pressure. This inverse proportionality comes from D in Equation 9 being inversely proportional to pressure and the partial pressure driving force in Equation 4 being constant. The maximum achievable rate of freeze-drying occurs at a pressure of 8 to 15 mm Hg, and the rate for helium is higher than that for nitrogen since both the thermal conductivity and the diffusivity with helium are higher than with nitrogen. Similar analyses were made by Harper[184] who recommended using helium at 0.5 to 10 mm Hg somehow in freeze-drying processes, and by Kan and deWinter.[33]

Carrier Gas

Means of employing helium effectively in a large-scale freeze-drying process at 2 to 7 mm Hg pressure were considered by Kan and deWinter.[33] Their first approach was simply to introduce helium into a standard batch tray-type freeze-dryer. They found, however, that the condenser surface would have to be placed within a very few cm of the food surface if the drying rate was not to be excessively lowered

FIGURE 14

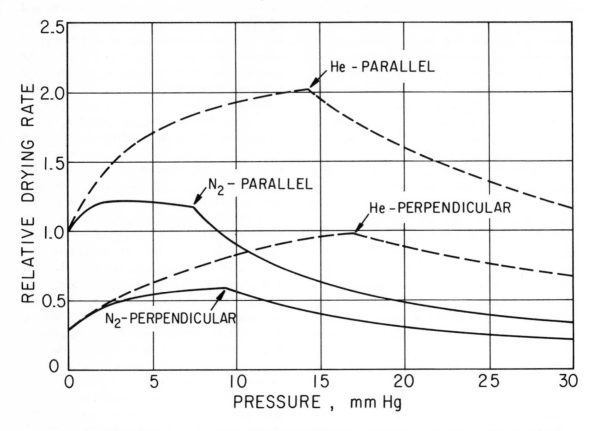

Relative freeze-drying rates vs. total pressure for freeze-drying turkey breast meat in helium and in nitrogen. Results are shown for drying parallel to the meat fiber and perpendicular to the meat fiber. Maximum outer surface temperature = 60°C. Maximum frozen zone temperature = −5°C.

by the increased diffusional resistance of the gas between the food surface and the condenser. A complex dryer geometry and condenser system would be required to overcome this limitation. The second approach considered by Kan and deWinter[33,185] was to circulate the gas in the drying chamber in order to remove the water vapor from the food surface. For a total pressure of 4 mm Hg they found that a gas circulation rate of 79 cfm/ft² of tray area was necessary to remove the water vapor generated by a drying rate of a modest 1 lb H_2O/hr ft², even if the gas leaving the chamber contained 2 mm Hg partial pressure of water vapor. A very large fan would be required for the circulation.

The two processes using helium outlined by Kan and deWinter[33] use steam-heated radiators for heat supply. Because of the low volumetric heat capacity of a low-pressure gas, much

more helium at 4 mm Hg, about 4800 cfm/ft² for 1 lb/hr ft² water removal, would be required to supply the heat of sublimation entirely by convection from the circulating gas.

Larson, Steinberg and Nelson[186] evaluated the influence upon freeze-drying rates for beef caused by the addition of various gases to the drying chamber. The effect found was small, probably because the gases were added in such a way that they would not be present to a large enough extent near the food material.

There have been other approaches toward reducing the gas compression requirements associated with a process in which the water vapor is to be removed by convection in another gas. Thuse[187] proposes using a circulating carrier gas only toward the end of a freeze-drying cycle. Barth, Pelmulder, Thuse and Blake[188–190] have devised a process employing a circulating condensable gas, such as heptane or a moderately heavy Freon, for water vapor removal. The vapor mixture is condensed upon leaving the drying chamber or within the drying chamber, the ice is separated from the liquid formed from the condensed gas, and the liquid is pumped to a higher pressure, revaporized and recirculated to the drying chamber. Since the carrier gas is condensable in this case, it is possible to avoid the gas compression step of other carrier-gas processes, but the circulation rate is still large, and the process may present phase separation problems and give odor contamination.

Fluidized Beds

Others have endeavored to combine the use of an inert gas with the use of a fluidized bed for freeze-drying. Mink, Sachsel, Dryden and Nack[191-194] have explored fluidized-bed freeze-drying processes with and without other solid particles being added to improve the fluidization and/or the heat transfer. Fluidization was accomplished by drawing a vacuum at the top of the bed and admitting air or another gas into the chamber below the bed. The times required for freeze-drying of various particulate foodstuffs[194] were relatively short compared to times required for radiant-heat freeze-drying in a tray dryer. Two difficulties were found which limited the fluidized bed heights which could be employed to a maximum of a few inches. The first limitation was the pressure build-up at the bottom of the bed, caused by the pressure drop necessary to create the fluidization. Too high a pressure at the bottom of the bed would cause melting. The second limitation came from the expansion of the gas as the pressure decreased going upward in the bed and resulted in only the top few inches being fluidized. These constraints limited the food loading to about 0.7 lb/ft² of bed cross-sectional area. Malecki et al.[80] endeavored to use fluidized beds for the freeze-drying of apple juice particles in a cold gas at atmospheric pressure but found it necessary to hold the bed temperature at $-34°C$, or less, in order to prevent the particles from sticking together. At these low temperatures and with the diffusional resistances encountered for atmospheric pressure, the rates of freeze-drying were very low, on the order of

1% water removal per hour.

Cycled Pressure

The thermal conductivity of the dry layer increases with increasing pressure while the effective diffusivity of the dried layer increases with decreasing pressure. At the optimum pressure found by the type of analysis which leads to Figure 14, neither the thermal conductivity nor the effective diffusivity will be at the maximum possible value; yet increasing one will decrease the other sufficiently to cause an overall reduction in rate. This compromise between thermal conductivity and effective diffusivity is necessary and results from the basic nature of the two transport processes. The transitions in both occur at pressures where the mean free path of the gas is comparable to the pore spacing within the dry layer. This compromise between thermal conductivity and effective diffusivity has led a number of engineers concerned with freeze-drying to contemplate a process in which the total pressure is regularly cycled up and down, between a high pressure where the thermal conductivity would be high and a low pressure where the effective diffusivity would be high. The heat input would be accelerated at the higher pressure and the removal of water vapor would be accelerated at the lower pressure, with the heat capacity of the frozen zone absorbing the differences. The temperature of the frozen zone would increase during the period of higher pressure and decrease during the period of lower pressure. Patents on processes of this sort have been issued to deBuhr,[195] Kan,[196] and Mellor.[197] deBuhr[195] suggests the use of hydrogen, helium or various other inert gases with the pressure cycled between 4 and 8 mm Hg and with a cycle time on the order of minutes. deBuhr indicates reductions of drying time in excess of 30%. Mellor[197] suggests the use of air or helium and cites experimental examples in a batch tray dryer where pressure cycle times were on the order of minutes and the pressure was cycled between 0.4 and 20 mm Hg, with the frozen zone temperature varying over a range of 7°F. Decreases in drying time by 25 to 40% as compared to ordinary freeze-drying at the lower pressure were found.

In analyzing the potential of such a cycled

pressure process, it is important to separate the effects of cycling from the effects of the higher pressure itself. Figure 14, covering operation at constant total pressure, shows that considerable reductions in drying time are possible at intermediate pressures as compared to operation at very low pressures. Although the heat input may be relatively fast during the high-pressure portion of a cyclic process, it will be relatively low during the low-pressure portion of the cycle. The net rate of heat input comes from an integration over a complete cycle, as follows:

$$q = \frac{1}{t_c \Delta L} \int_0^{t_c} k(T_s - T_f) \, dt , \qquad (25)$$

where t_c is the time for a complete cycle (high pressure + low pressure). The comparable equation for steady-state operation is obtained by combining Equations 2 and 8:

$$q = \frac{k}{\Delta L} (T_s - T_f). \qquad (26)$$

In Equation 25 k is a function of pressure (which is, in turn, a function of time) and T_f is a function of time. For steady-state operation, a particular value of T_f will be associated with each value of pressure and, hence, with each value of k. In the cycled operation T_f must vary between the limits of the two steady-state values of T_f corresponding to the high and low pressure level but will lag behind the pressure because of the heat capacity of the frozen material. The average value of $T_s - T_f$ during either half of the cycle will correspond very nearly to the value of T_f lying halfway between the extreme values. The average thermal conductivity will lie somewhere between the two extreme values, probably displaced toward a higher value than the average because of the high pressure portion of the cycle being longer than the low pressure portion.

From this analysis it can be seen that cycling of the chamber pressure over quite a wide compression ratio would be required to gain even a small percentage improvement in the overall drying rate due to cycling alone. Using the thermal conductivity vs. pressure curves for pear obtained by Harper and El-Sahrigi[53] and presented in Figure 6 as an example, the peak attainable drying rate at constant pressure might occur at 10 mm Hg chamber pressure. For helium the thermal conductivity is raised by 27% and 66% above the 10 mm Hg value at pressures of 20 mm Hg and 100 mm Hg, respectively. For nitrogen these two percentages become 8% and 15%. Most of these gains in thermal conductivity would be eradicated by the need for operating at a pressure well below 10 mm Hg during the low-pressure portion of the cycle in order to bring T_f back down enough so as to avoid melting in the next high-pressure period. The thermal conductivity in the low-pressure period would be reduced by a percentage comparable to the gain during the high-pressure period. The only salvation giving any net gain is a slightly greater $T_s - T_f$ driving force on the average and the exponential vapor pressure-temperature relationship, which will make the high pressure portion of the cycle longer than the low pressure portion. Even so, one can safely conclude that the benefit caused by cycling as opposed to operation at a constant, optimum pressure is virtually nonexistent for the nitrogen case, and might be as high as a 10 to 15% overall increase in rate in the helium case if the pressure is cycled over a compression ratio of 200 or more. This is hardly enough to warrant the complication and expense of cycled pressure.

Cycling the pressure will also serve to remove the water vapor by entrainment in the gas removed during the pump-down portions of the cycle; however, the gas usage or circulation requirements are the same as those required in the Kan and deWinter process for circulating gas at constant pressure. The cycling of pressure will also provide agitation and circulation patterns within the gas space of the chamber which can promote external mass and heat transfer coefficients, but there are easier ways of accomplishing the same effect. Cycled pressure freeze-drying is therefore not an attractive process.

Convective Freeze-Drying in Layered or Mixed Beds

The basic limitation on a convective freeze-drying process is the very low mass capacity of a gas which can take up water vapor to a partial pressure of about 2 mm Hg, at most. There is a still greater limitation to heat supply by convection because of the very low volumetric heat capacity of a gas circulating at, say, 15 mm Hg total pressure. King and Clark[25,26,198] have proposed a convective freeze-drying process which circumvents these limitations by reusing the circulating gas over and over again. As shown in Figure 15 for the layered-bed version, a circulating light gas, such as helium or nitrogen, passes successively through alternating layers of food pieces and desiccant particles. In passing through a food layer, the gas loses heat to the food, causing sublimation, and, at the same time, picks up the generated water vapor. Then, in passing through the next layer of desiccant, the gas

FIGURE 15

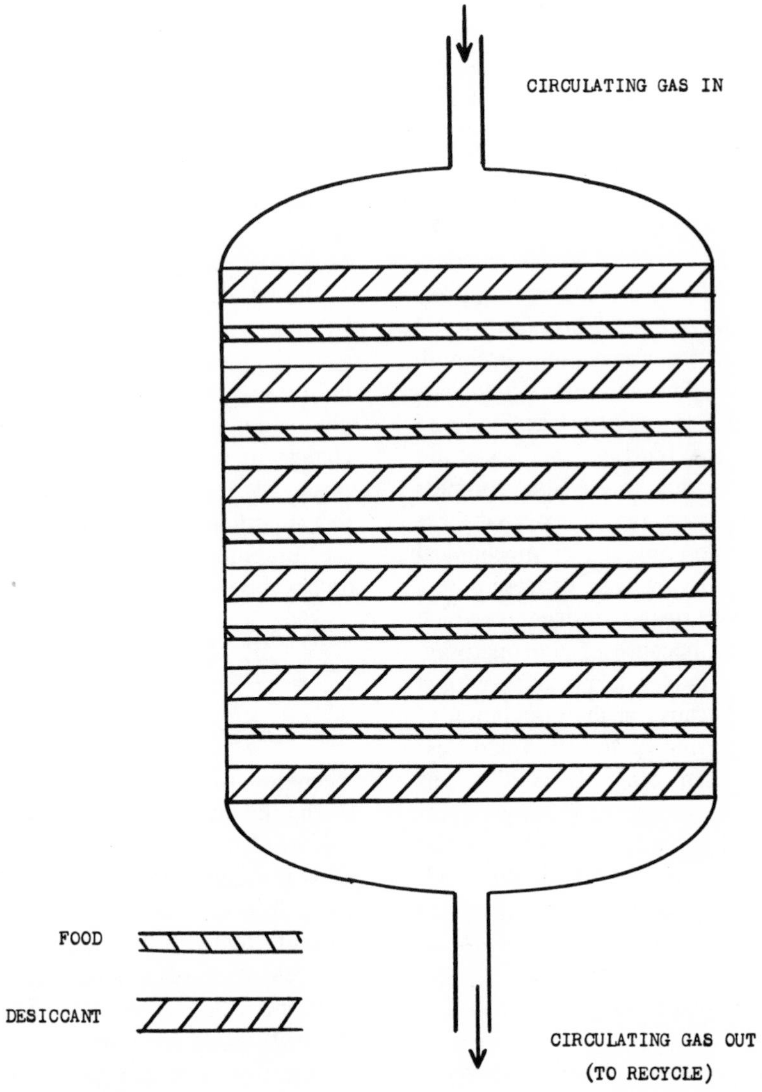

CIRCULATING GAS IN

FOOD

DESICCANT

CIRCULATING GAS OUT
(TO RECYCLE)

Freeze-drying using a light gas circulating at moderate pressure through alternating layers of food and desiccant, after King and Clark.[26,198]

loses the water vapor to the desiccant and is dried out so as to be readied to remove water vapor from the next layer of food pieces. Also, and most important, the gas is reheated in the desiccant layer by the heat of sorption of the water vapor taken up by the desiccant. The heat of sorption per lb of water on most desiccants is nearly equal, and usually greater than, the heat required for sublimation per lb of water in the food. Consequently, the desiccant itself serves as the principal heat source for freeze-drying, with only a trim heater and/or cooler being required in the gas circulation line external to the drying chamber. This trim heat exchanger ensures that the gas issuing from the desiccant layers is below the temperature at which there would be thermal damage to the dried outer surface of the food in the next layer (about 60°C in most cases) and below the temperature which would cause melting in the next layer.

Because of the wet-bulb-like temperature depression accompanying freeze-drying, the gas can be at a temperature above ambient without melting the food, as long as the pressure is not too high. A typical pressure would be near or somewhat above the optimum shown in Figure 14. The gas serves as a carrier of heat from the desiccant to the food and of moisture from the food to the desiccant. Primary features of this layered-bed process are lack of need for a major external heat source or for a major amount of mechanical refrigeration during drying. The process differs from other processes using desiccants which have been suggested because of the operation of the process at a moderate pressure with a light inert gas and because of the circulation of the gas and interspersion of the food and desiccant. It differs from other circulating gas processes in that the drying capacity per cfm of gas circulation is greatly increased by the use of a desiccant interspersed with the food. If N layers of food material alternating with desiccant layers are used, the gas circulation requirement is 1/N of that required if the desiccant were not placed in the intervening layers.

The moisture capacity of the desiccant layers must be such that the loaded desiccant can still remove moisture from the food

toward the end of freeze-drying. Molecular sieve has good characteristics in this respect, but still the amount of desiccant required is about ten times the original weight of food. Regeneration of the molecular sieve desiccant would be accomplished by heat from a relatively inexpensive source, such as combustion flue gases, at the end of drying. Molecular sieve particles are regenerated and reused 3000 times and more before replacement in most services, and that should be the case in this process because of the isolation of the desiccant from the food and the lack of handling of the desiccant.

In the layered-bed process the molecular sieve can remain as layers in the drying chamber, and the food can be inserted and removed on screen-bottom trays fitting in between the desiccant layers. Clark and King[25,26,198] also point out that direct mixing of the food and molecular sieve gives even more efficient drying, although separation of the food from the desiccant will be required after drying. Molecular sieve spheres do not erode easily, so it should be possible to avoid contamination by fines.

In the analysis of the early portion of drying in the layered-bed process it is important to give attention to the maximum allowable change in temperature of the gas as it passes through a food layer.[26] For a given drying rate the superficial gas velocity will be related to the pressure and to the temperature drop across the layer as follows:

$$u_G = \frac{\Delta H_s \rho_L w_o RT}{P\, C_p (\Delta T)_L} \left(- \frac{dX}{d\theta} \right) \qquad (27)$$

where u_G is the superficial gas velocity, ΔH_s is the latent heat of sublimation, ρ_L is the loading density of food in a layer (i.e., lb/ft²), w_o is the initial weight fraction of water in the food, R is the gas constant, T is the average absolute temperature, P is pressure, C_p is the heat capacity of the gas, $(\Delta T)_L$ is the temperature drop across one layer of food, X is the fraction of the initial water present, and θ is time. For a temperature swing of 50°C and a loading density of 1 lb/ft², a helium superficial velocity of 4 ft/sec at 16 mm Hg will give an initial drying rate of 40% water removal per hour. The required u_G will be inversely proportional to tray loading, directly proportional to the desired initial drying rate and relatively independent of pressure.[26] The number of layers used can be large, subject to the allowable pressure drop of the gas passing through the drying chamber.

Clark and King[25,26] report the results of bench-scale tests of the mixed-bed version of this process and give a mathematical model for analyzing the rates of freeze-drying attainable. For 3/8-inch cubes of turkey meat freeze-drying to a moisture content under 3% was attained within three hours with a circulating mixture of helium and nitrogen.

Freeze-Drying with Added Liquids

Another procedure which has been suggested for supplying heat and removing the water vapor generated in freeze-drying involves immersing the substance to be dried directly into a liquid boiling under vacuum.[199,200] If the boiling temperature of the liquid is held at an appropriate value by adjusting the pressure of the system, heat will be transferred to the material to be dried, and the drying rate will be sufficient to keep the frozen zone of the material being dried from melting. So that diffusion within the outer shell of the material being dried will not be a serious rate-limiting factor, the boiling liquid should generate an equilibrium vapor pressure in the range of the optimum pressures indicated in Figure 14, at the boiling temperature which is used. Such a process may be looked upon as being related to the condensable carrier gas freeze-drying process of Barth et al.[188-190] which was described earlier, the difference being that in the present case the carrier is boiled directly adjacent to the material being dried. The process has some similarities to azeotropic distillation; hence, the name "azeotropic freeze-drying" has been used.

The concept of azeotropic freeze-drying was apparently first suggested by Wistreich and Blake,[199] who freeze-dried frozen ground meat by immersion into toluene boiling at 30°C and 22 mm Hg absolute pressure. They indicated that there appeared to be no melting of the sample during drying. Bohrer[200] felt that it was more desirable to use a liquid which is less harmful in residual amounts in foods and which forms an azeotrope with water while being at least partially miscible with water. Bohrer recommended ethyl and n-propyl alcohols and acetates and reports drying of hamburger in an ethyl acetate-water azeotrope boiling at 24°C and 100 mm Hg. It seems questionable that the meat would remain frozen for drying at such a high pressure with the heat source at 24°C.

The liquid used in an azeotropic freeze-drying process would permeate the foodstuff being freeze-dried to at least some extent. Even if a high-temperature vacuum operation were used to remove residual liquid after drying, there would still be concern that some of the liquid would remain, contributing to taste and odor.

Rey[4,201] has suggested sublimation of solvents other than water for the stabilization of biological species. Specific solvents suggested are ammonia, carbon dioxide, and low-melting halogenated hydrocarbons and aliphatic amines. For example, 1-lysine can be dissolved in liquid ammonia, the solution can be frozen and the ammonia can be removed by sublimation under vacuum at a temperature of about −110°C. This approach is suited to biological species which are insoluble in water or which are unstable at the temperatures normally used for ice-sublimation freeze-drying.

Microwave Heating

Very rapid rates of freeze-drying, comparatively, would be possible if heat could somehow be supplied to the sublimation front fast enough to maintain the maximum allowable sublimation front temperature when a very low pressure is maintained in the drying chamber. At a very low pressure D' in Equation 9 would have a high value reflecting primarily the contribution of Knudsen flow. k_{gi} would therefore be as high as possible. k_{ge} would also be high because of the very low pressure, and for the fixed, maximum possible, water vapor partial pressure driving force corresponding to the maximum allowable frozen zone temperature, the rate of drying (N_A) would be very fast, at least an order of magnitude faster than in most actual freeze-drying processes.

The factor which prevents this happy combination of very high mass transfer coefficient and maximum allowable partial pressure difference driving force from occurring in freeze-drying processes is the difficulty of supplying heat at the necessary rate to the sublimation front in a system at very low pressure. Rates of heat conduction provided by the low thermal conductivities of freeze-dried foods are much too low. Even for heating by conduction from one side across a frozen layer there is a significant resistance to heat conduction. Also, to achieve reasonable capacity without excessive constructional complexity in a process where heat is conducted through the frozen layer it is necessary to freeze-dry rather thick slabs of material. The increased material thickness results in longer vapor diffusion paths within the dry layer, reducing the drying rate.

Microwave heating offers the possibility of delivering the heat of sublimation directly to the sublimation zone, thereby overcoming the problems of supplying heat by conduction from the surface of the material being dried. Microwave energy is dissipated and generates heat in regions of high dielectric constant or, more correctly, high loss factor. Water has a very high loss factor and a high dielectric constant, much higher than the values for dried foods. Consequently, in a microwave field containing a partially dried foodstuff the microwave energy will be dissipated and cause heating in the region of the food where the water is. If microwave energy could be dissipated rapidly enough in the frozen core of a food undergoing freeze-drying, it might be possible to achieve the ideal combination of a very high internal mass transfer coefficient from low pressure along with the maximum allowable water vapor partial pressure difference driving force, corresponding to the maximum allowable temperature of the sublimation front.

The theory and practice of microwave heating have been detailed elsewhere, for example, by Copson,[202] and will not be repeated here. Early experiments with microwave heating for freeze-drying were reported by Harper, Chichester and Roberts,[203] by Decareau,[204] by Copson,[202] and by Horejsi and Fric,[205] among others. The problems encountered have been outlined by Meryman[206] and by Harper et al.[203] as follows:

Ionization—When the microwave energy creates sufficient gas ionization to cause a glow discharge or arcing near or within the dry layer of the food, discoloration and flavor damage to the food can result. In addition, the discharge consumes a large amount of microwave energy.

Uneven Heating—Since the amount of energy dissipated per unit volume of water is roughly uniform in the drying chamber, a nonuniformity of piece sizes of the material being dried will cause piece-to-piece variations in the temperature of the frozen zone. Increasing temperatures give higher loss factors for most materials and, hence, there is a potential instability in which locally warmer regions of the material absorb heat faster and thereby become even more overheated. Finally, fat in the dry portion of a meat will retain a significant loss factor and may thus be overheated.

Impedance Matching—The impedance of the contents of the drying chamber must be matched with the output of the microwave generator. The chamber impedance changes continuously during drying, making adjustments necessary.

Low Loss Factor for Ice—Although water has a high dielectric constant and a high loss factor, ice has a very low dielectric constant and a very low loss factor,[204] as has been confirmed by microwave heating experiments with pure ice.[207] Consequently, the microwave energy is dissipated almost entirely to the bound, or unfreezable, water in freeze-drying. This fact limits the rate at which microwave energy can be absorbed by frozen foodstuffs and may also cause energy dissipation in the dry layer, to the extent that bound water is left behind by the retreating sublimation front.

Hoover, Markantonatos and Parker[208,209] carried out extensive tests of freeze-drying of a number of different foods in a microwave freeze-dryer. Among their findings were that the drying times required for different thicknesses of foods (such as beef patties ranging from 0.25 to 1.0 inch thick) were independent of thickness and were rather constant for different foods at values in the vicinity of 2.5 hours. As has been suggested by Gouigo et al..[207] these results indicate that the drying process was rate-limited by the ability of the device to dissipate heat within the food pieces. This, in turn, suggests that the intensity of the microwave energy supply was limiting. Hoover et al.[208] found that the intensity of microwave heating which they could use was limited not by the generating device but, instead, by the need for avoiding a glow discharge. The relationship for "glow power" vs. chamber pressure which they found is shown in Figure 16.

FIGURE 16

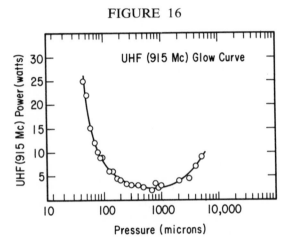

Glow power vs. chamber pressure for microwave freeze-drying, after Hoover, Markantonatos and Parker.[208] Copyright 1966 by Institute of Food Technologists.

The glow power is the minimum power necessary to cause glow discharge within the drying chamber, in this case for a fresh frozen beef patty on a Teflon platform. The glow-power curve will vary from one microwave introduction scheme to another. Hoover et al.[208] used chamber pressures of 20 to 60 microns typically and used a maximum power input of 15 watts. Even with this limitation, the time reported by Hoover et al. for freeze-drying pieces of foods of a given size are considerably shorter than those which are required for freeze-drying in conventional processes.

From Figure 16 it should be noted that the glow discharge limitation to input power is most severe in the very range of pressures which are most commonly used for freeze-drying (500 microns to 1 mm Hg).

Hammond[210] used the results found by Hoover et al.[208,209] to evaluate the economic incentive for microwave freeze-drying as opposed to ordinary radiant-heat, tray freeze-drying. The microwave process has a much shorter cycle time, which decreases the loading capacity required for a unit to produce a given amount of product per unit time. As a result, the capital investment predicted for a microwave plant of a given annual production capacity was less than that for a conventional plant. On the other hand, the energy input rate is related to the average rate of water removal, which is the same for the two plants. Even though Hoover et al. found microwave energy utilization efficiencies of 65 to 90% over most of the drying cycle, the energy for a microwave process is relatively expensive electrical energy. This factor, coupled with the need for tube replacement, makes the operating cost of a microwave plant about 50% greater than that for a conventional plant. Putting the lower capital investment together with the greater operating cost, Hammond finds the real costs of the two processes to be nearly the same. He also points out the incentive for a processing approach which combines conventional freeze-drying for the removal of the first 60 to 80% of the water with microwave freeze-drying for removal of the remaining water and shows that the cost of the combined process should be less than that of either process alone. The opinion that microwave freeze-drying will find its greatest incentive for terminal freeze-drying has been voiced by numerous others and is probably correct. Decareau[204] earlier made cost predictions for the two processes and the combined process which are similar to those of Hammond,[210] but the promise evidenced by these figures for microwave heating has yet to be shown in practice. The glow discharge and thermal instability problems still inhibit the use of microwave heating. Also, the level of vacuum required to avoid glow discharge at reasonable power inputs is not always easy to attain.

Gouigo, Malkov and Kaoukhtchechvili[207] report the results of microwave freeze-drying at 2450 Mc (as opposed to 600 to 900 Mc used by Hoover et al.), with a chamber pressure of 60 to 70 microns. The food studied was slices of beef measuring 30 x 60 mm and 6 to 25 mm thick. They report being able to use high power inputs from a 150-watt maximum power generator without gas ionization because of the design chosen for their drying process and its location within the microwave field. Gouigo et al.[207] report freeze-drying times of 60 minutes for 25-mm-thick beef slices, and 20 to 30 minutes for slices 6 mm thick. The time for slices 2 to 3 mm thick was less than 15 minutes. They take the fact that the drying time increased markedly with sample thickness to indicate that the rate-limiting factor was not the rate at which microwave energy could be dissipated but, instead, internal mass transfer resistance due to the outer dry layer. This claim was reportedly verified by careful pressure measurements made by means of pressure probes located within the pieces being dried, at the piece surface, and in the drying chamber. Apparently a pressure gradient was found within the pieces, but not in the drying chamber. For complete internal mass transfer control one would expect a still greater dependence of drying time upon sample thickness, however, since Equation 14 and the mass transfer analogs of Equations 16 and 17 show that for internal mass transfer control the drying time should vary as the square of the average piece dimension. A freeze-drying process limited by internal mass transfer at chamber pressures on the order of 65 microns represents the maximum attainable rate of freeze-drying with good heat input, referred to above.

Through measurements with imbedded thermocouples Gouigo et al.[207] found that the temperature of the center of a piece was appreciably higher than the temperature of the sublimation front and, in fact, found that the temperature of the center of a piece could even approach 0°C. They indicate that the centers of the pieces were actually melted but that the drying was fast enough to surround the melted core with a solid layer of ice. Thus the sublimation would still occur from a retreating ice front which would preserve the structure at the point of drying, but the center of the piece would be unfrozen. The unfrozen water in the central core would provide the very high loss factor characteristic of liquid water and thus scavenge the microwave energy to a much greater extent per unit volume than would be the case for a frozen core.

The concept of picking up microwave energy with a liquid-water core and supplying that energy by conduction from that core to a sublimation front which is kept frozen by the very rapid sublimation rate is appealing, but it would seem difficult to achieve this behavior in some pieces within a drying chamber without melting other pieces to an extent great enough to cause a poorer product. On the other hand, a freeze-drying process which can dry ¼-inch beef slabs in 25 minutes may be able to stand lesser product quality and still be economically competitive.

Heating with Infrared Radiation

Microwave heating is one form of radiant heating, with the particular feature of being able to penetrate the food material to the point where the moisture is located. Other forms of radiant heating can also give these penetrating characteristics to some extent. Ginzburg and Lyakhovitsky[211] have reported the use of penetrating infrared radiation for freeze-drying. In their process the radiant heat is supplied through the outer dry layer of the food material. The wavelength of the radiation is selected so as to give a high ice absorptivity along with a high permeability through the dry layer. They report freeze-drying conditions wherein a radiation wavelength of 1.3 microns was used from a filament temperature of 1500 to 1800°C with an energy flux density of 0.2 to 0.5 watts/cm² from the heating panels. In an earlier paper Zamzow and Marshall[212] tested and analyzed the use of penetrating radiation to supply heat to the sublimation front through the frozen layer in freeze-drying of liquids.

Freeze-Drying of Liquid Foodstuffs

Freeze-drying of frozen liquid foodstuffs has usually been accomplished industrially in batch, tray-type units of the type used for particulate solid foods. The liquid is frozen and then granulated to be put on the trays. One of the reasons for handling liquids in this way is that most large freeze-dryers to date have been multi-product freeze-dryers, wherein particulate solids and frozen liquids might both be handled at different times. Liquids have the particular feature of not being restricted in particle shape, and it should be possible to take advantage of this characteristic in order to devise improved freeze-drying processing approaches which are specific for liquid foodstuffs.

Greaves[213] and Hackenberg[214] have suggested continually abrading the dry layer from a substance during freeze-drying by means of a scraper or a rotating wire brush. Such a process reduces the internal resistances to heat and mass transfer but produces a product of small particles which may be undesirable or which may entrain in the escaping water vapor.

There have been two processes recently introduced which involve freezing and, subsequently, freeze-drying a liquid foodstuff on the surfaces of vertical tubes. These processes are reviewed by Bengtsson.[140,141] The Seffinga process[215,216] is used commercially and involves flowing the liquid feed into the tube interiors of a bank of vertical tubes which are chilled by condensing ammonia refrigerant in the tube jacket. A layer of frozen substances 7 to 10 mm thick forms on the insides of the tubes. The excess liquid feed is then drained, and freeze-drying proceeds with higher pressure ammonia delivering the heat of sublimation by condensing outside the tubes. The freeze-dried product should fall readily from the interior tube surface after drying while having maintained good thermal contact with the surface during drying. It is possible that the impermeable surface layer observed by Quest and Karel[67] and others, as noted above, may hamper drying. A similar device in which the liquid freezes and dries on the outer surfaces of a bank of vertical tubes has been built in pilot scale by Mitchell-Edwards.[217] In this device a circulating brine serves as a cooling medium for freezing and a heating medium for drying.

Spray drying is widely used for the high-temperature drying of liquids, and it would therefore seem appropriate to explore the use of spray-freeze-drying for liquid foodstuffs. The first difficulty which one encounters is that an ordinary spray dryer achieves heat input from the sensible heat content of a large volume of heated gas. In a vacuum system where the droplet temperature must be held low enough to maintain a frozen condition this source of heat input is impractical. Hence, efforts to employ a spray for freeze-drying have turned to the use of the spray as a freezer to form small frozen particles, which can then be dried by other means.

Greaves[218] was one of the first to investigate spray-freezing, accomplished through cooling by evaporation of a portion of the liquid in the drops leaving the spraying device. He found that spraying into a high vacuum gave rapid evaporation and therefore rapid freezing, which could lead to solidification so near the spray device that the particles would be flake shaped or would block the nozzle. Maintaining a "poorish" vacuum alleviated this situation. Subsequent freeze-drying was to be accomplished in a tunnel through which a suspension of particles in the vapor phase would flow, but velocities in the tunnel were too high to give adequate residence times for drying. Greaves then turned to a device which exuded liquid to be frozen as a strip on a rotating plate, with particles being formed from the strip by a scraper.

Thuse et al.[219] have investigated a spray-freezer in which the liquid is sprayed upward into a vacuum chamber and solidifies into small particles through evaporative freezing. The particles are then transported continuously through a freeze-drying tunnel in a succession of pans. Niro Atomizer Co.[140,141] has developed a process wherein spray-freezing is accomplished by spraying into a very cold air stream ($-70°C$) or by evaporative freezing in a vacuum, with subsequent freeze-drying occurring in a rotating wire mesh drum surrounded by heating elements. Malecki et al.[80,220] have investigated an atmospheric pressure process wherein the liquid is frozen by spraying into a stream of cold gas and is subsequently freeze-dried in a fluidized bed. A similar process of spraying with evaporative freezing followed by fluidized drying in a vacuum is the approach of Mink et al.[191–194] described earlier. The atmospheric process has the drawback of low diffusion rates, while the vacuum process has the drawback of low bed heights. Hutton[221] suggests partial dehydration through ordinary spray drying, followed by freeze-drying of the resultant particles.

Problems encountered during vacuum spray-freezing are the loss of volatile flavor and aroma compounds through evaporation and the possibility of droplets sticking to the chamber wall. One approach to overcoming these difficulties is freezing through spraying directly into liquid nitrogen.[220]

Freeze-drying of liquids has the virtue of good aroma retention but requires low temperatures of the frozen zone to secure sufficient freezing for substances of high carbohydrate content. As a result, the driving force for mass transfer is low and drying rates are constrained. Drying from the liquid state gives much higher rates of drying but creates the problem of aroma loss. Chandrasekaran and King[86] find that water removal from the partially frozen state (a "slush", in which 20 to 70% of the water is solidified) through combined evaporation and sublimation can occur with nearly the same degree of aroma retention as occurs in freeze-drying. Because of the higher temperatures corresponding to the two-phase liquid and solid mixture, the drying rates achievable for slush-drying are higher than those for freeze-drying.

Pre-Concentration

Liquid foodstuffs also afford the opportunity for pre-concentration before freeze-drying through removal of part or most of the water by some other means. There are two major potential advantages to be gained from such a pre-concentration step. First of all, it is likely that the cost of the pre-concentration step per lb of water removed will be markedly less than the cost of freeze-drying per lb of water removed. Therefore, a combined process may be able to give the quality of a fully freeze-dried product at a lower cost. The second potential advantage lies in the known better aroma retention achieved in freeze-drying from a higher initial dissolved solids content, as discussed above. If the pre-concentration can be accomplished without a major aroma loss, then the aroma retention characteristics of the combined process can be as good as or even better than the aroma retention characteristics of freeze-drying by itself. Primary candidates for a pre-concentration process are freeze-concentration[222] and reverse osmosis.[223]

Elerath and Pitchon[82] have indicated that a preliminary freeze-concentration of coffee before freeze-drying improves the color of the product. It should also be noted that their two-step freezing process before freeze-drying is also in a way a preliminary concentration of the solids content of the liquid left after the first freezing stage, which should lead to improved aroma retention. Similarly, the slush-drying process put forth by Chandrasekaran and King may be looked upon as a preliminary freeze-concentration before evaporation or drying of the concentrate from the liquid state, thus giving improved aroma retention. The effect of the amount of pre-concentration upon the subsequent rate of freeze-drying has been investigated by Spiess et al.[60] for maté extract, by Sharon and Berk for tomato juice,[51] and by Monzini and Maltini[68] for orange juice. All have found that the maximum rate of product output from the freeze-dryer comes from a substantial degree of pre-concentration.

Davis and Pfluger[224] and others have suggested what is effectively a pre-concentration step for solid foodstuffs by using air drying followed by freeze-drying.

QUALITY CONTROL

The problem of maintaining a uniform and reliable high product quality in freeze-drying is an important challenge which has not yet been solved in a wholly satisfactory fashion. Aspects of quality control include (1) maintenance of uniform drying conditions for all pieces or particles so as to avoid uneven drying rates, (2) maintenance of as high as possible a rate of freeze-drying without product spoilage, and (3) prompt recognition of the end point of drying. General reviews of approaches to quality control have been given by Rowe,[5] Burke and Decareau,[10] and Havelka.[225]

Piece-to-Piece Scatter

Clark[24] reports the terminal moisture contents of 98 different ½-inch cubes of turkey meat freeze-dried for four hours in a tray-type, batch freeze-dryer with the piece surface temperature held near 150°F and with air admitted to maintain a 2-mm-Hg total chamber pressure despite a low condenser temperature. Conditions were purposefully chosen to give less-than-complete drying. The results are shown in Figure 17, from which it is apparent that considerable differences in drying rate occur for different pieces in this type of freeze-dryer. Kan and deWinter[33] also report uneven drying in their process for freeze-drying in helium.

Variations in terminal moisture content from piece-to-piece can occur for a number of reasons, including (1) differences in piece size, which would be reflected by the influence of L^2 in drying rate expressions, such as Equations 12 through 15, (2) uneven exposure to the heat source, (3) variations from piece to piece in thermal conductivity, such as were found for turkey meat by Triebes and King,[32] and (4) local build-ups in the water vapor partial pressure in the chamber, which would alter the mass transfer driving force.

Maintenance of Maximum Rate of Freeze-Drying

In order to maintain the maximum possible rate of freeze-drying, it is desirable to monitor the two possible limiting factors, the tempera-

ture of the frozen zone and the temperature of the outer dry surface, continuously. Although one approach has been to insert a number of thermocouples into the drying chamber for exactly this purpose, a complexity stems from the impossibility of making such measurements for each and every piece being freeze-dried. The temperatures may be maintained correctly for the piece or pieces being monitored, but this does not guarantee that other pieces are not being overheated.

One promising approach for process monitoring, which has been developed and reported by Bouldoires,[226] is the continuous measurement of the dielectric constant of the bulk of the material being dried. Since liquid water has a much higher dielectric constant than ice, dried food or other materials which may be in the chamber, it is possible to detect quite rapidly any onset of melting of the frozen zone in any piece within the field being monitored so that corrective action may be taken promptly.

For conditions of freeze-drying where the maximum temperature of the frozen zone is a limitation, such as for carbohydrate-containing food liquids, Rey[D,227,228] has developed a procedure for monitoring the electrical resistance of the frozen zone. Since good crystallization

FIGURE 17

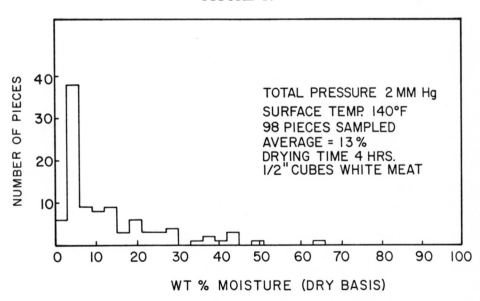

Piece-to-piece variation of final moisture content in freeze-dried turkey, after Clark.[24]

71

and electrical resistance seem to be related,[D] this approach provides the maximum drying rate consistent with adequate crystallization of the frozen zone but again applies only to the particular locations being monitored.

Neumann and co-workers[183, 229-231] in the Leybold processes have made use of a technique of periodically blocking off the condenser and observing the build-up of pressure within the drying chamber. Because of the heat capacity of the frozen material this pressure measurement can amount to a measurement of the temperature of the frozen zone for those particles with the highest frozen zone temperatures. This method has the benefit of giving a measurement characterizing the pieces nearest to the drying limit but suffers if inert gases are present.

Bouldoires[226] has reported on the use of a continuous-reading vapor hygrometer which is insensitive to vacuum and which will, therefore, continuously record the dew point of the water vapor in the drying chamber. The dew point is directly related to the partial pressure of water vapor and, thus, can indicate indirectly the rate at which water vapor is being removed. Illich[232] has shown how the partial pressure of water vapor in the chamber can be measured through combined use of an ionization vacuum gage and a thermocouple vacuum gage. Flosdorf[1] reports a device suggested by Sherwood in which a ball of ice with an imbedded thermocouple is placed in the drying chamber, and the chamber humidity is recorded from the ice temperature through wet-bulb thermometry.

It should be noted that none of these methods measures the outer dry surface temperature of the pieces, which is the rate-limiting condition under conditions of heat-transfercontrolled freeze-drying. At present there appears to be no good substitute for monitoring surface temperatures at a number of locations through thermocouples or for relying upon operating experience.

End Point Determination

It is important to stop a freeze-drying process at the point where the material is adequately dried so as to avoid the product deterioration which can occur from overdrying. The measurements presented in the previous section can also give an indication of the completion of drying. For end point recognition it is preferable to measure characteristics representing all the food material

which is in the freeze-drying chamber. This can be accomplished through monitoring the dielectric constant of the field containing the drying charge,[226] by monitoring the water vapor partial pressure in the drying chamber[1, 226, 232] or by periodically measuring the pressure build-up when the condenser is blocked off.[183, 229-231]

Burke and Decareau[10] give an extensive survey of end point recognition approaches under investigation prior to 1964.

ADDITIONAL DEVELOPMENTS

This section covers additional developments which have come to the author's attention since the original review for *CRC Critical Reviews in Food Technology* was prepared.

Reviews

Ginnette and Kaufman[233] have given a comprehensive review of conventional processing equipment used for freeze-drying and have provided means for designing and sizing equipment. Brockmann[234] and Rothmayr[235] have reviewed the capabilities of freeze-drying and processing problems in a more qualitative fashion. A review of recent developments in low-temperature food dehydration processes, including freeze-drying, was given by King.[236]

Non-Food Uses

One of the original large-scale uses of freeze-drying was in connection with the stabilization of penicillin. Brockmann[234] has described the development of freeze-drying for that purpose during World War II.

Schnettler, Monforte, and Rhodes[237] have found freeze-drying to be effective as a starting point for the preparation of homogeneously mixed ceramics. Freeze-drying of solutions of mixed salts yields a porous solid structure which can then be converted to ceramic. Work on the use of this technique for preparing new ceramic materials is continuing.

Physical Properties

Jason and Jowitt[238] have prepared an index to available data on various physical properties of foodstuffs. Their coverage is quite extensive and includes a number of references to sources of properties relevant to freeze-drying.

Freezing Conditions

Riedel[239] has summarized the large number of studies which he has made over the years on calorimetric measurements of freezing phenomena in foods. Partmann[240] gives a complementary summary of chemical and biochemical changes occurring during, and because of, freezing. Huber, Harrington, and Stadelman[241] recently reported the influences of both freezing and freeze-drying upon the solubility of proteins in meats.

Aroma Retention

Experimental measurements of the retention of volatile organic species during freeze-drying have been reported by Sauvageot, Beley, Marchand, and Simatos[242] and by Flink and Karel.[243] The latter authors examined synthetic aqueous carbohydrate solutions. Sauvageot et al.[242] used flame ionization chromatography to monitor naturally-occurring volatile compounds during the freeze-drying of orange and raspberry juices and found that 80 to 90% retention of these compounds is possible under well-chosen conditions of freezing and freeze-drying.

It is now possible to compare the results of these and other experimental studies with the predictions of the selective diffusion concept of volatiles retention, originally set forward by Rulkens and Thijssen[F,89] and discussed at length earlier in the present review. Following and extending somewhat[236] the approach of Rulkens and Thijssen, a diffusion analysis shows that the extent of volatiles loss during freeze-drying should decrease with decreasing values of the dimensionless group (Dt/L^2), where D is the volatiles effective diffusivity, t is the time during which diffusion can occur, and L is the size or thickness of the region from which volatiles loss occurs. Table 3 shows three factors which should increase

TABLE 3

Factors Increasing Volatiles Retention
by Reducing the Group (Dt/L^2)

Factor	Effect
Greater Velocity of Sublimation Front (Faster Freeze-Drying Rate)	Lower t
Greater Thickness of Concentrate Microregions	Higher L
Greater Dissolved Solids Content of Concentrate Microregions	Lower D

volatiles retention in freeze-drying by decreasing the group Dt/L^2.

Following Table 3, the effects of a number of processing variables during freeze-drying can be predicted:

1. A *higher dissolved solids content before freezing* increases the volumetric ratio of concentrate to ice and thereby should increase L for the microregions of concentrate formed during freezing. By Table 3 the higher L should lead to improved volatiles retention. In cases of non-equilibrium freezing or for incomplete freezing a higher dissolved solids content before freezing may also lead to a higher dissolved solids content of the concentrate microregions formed during freezing, thereby reducing D and also promoting volatiles retention.

Higher volatiles retention for greater dissolved solids contents before freezing has been observed experimentally by Rey and Bastien,[D] Chandrasekaran and King,[86] Flink and Karel,[243] and Sauvageot et al.[242] although the last-named authors did not observe improved volatiles content at higher initial dissolved solids content for all the components which they monitored. Sauvageot et al.[242] further observed that there would be poorer volatiles retention if the initial dissolved solids content were so high that they encountered melting of the frozen core or collapse of the dry layer during freeze-drying. Flink and Karel[243] also found experimentally that increased volatiles loss would occur when drying conditions were such as to cause a collapse of the dry layer.

2. *Slower rates of freezing* lead to larger ice crystals and, hence, to thicker concentrate microregions. Sauvageot et al.[242] and Flink and Karel[243] both found a strong effect of slower freezing rates in giving improved volatiles retention.

3. *Thinner pieces* should give a higher velocity of the sublimation front and, hence, a lower t and increased volatiles retention, as long as the drying rate is controlled by internal resistances to heat and/or mass transfer. The critical time t for volatiles loss is the time of exposure of the newly-dried material while and just after the frozen front passes, not the total freezing-drying time required for the entire sample. Thus the important rate criterion for volatiles loss is the actual rate of passage of the sublimation front through the material (distance/time) and not the drying time itself.

Improved volatiles retention for smaller sample thickness has been found experimentally by Sauvageot et al.[242] and by Flink and Karel.[243] The latter investigators found that the beneficial effect upon volatiles retention from decreasing sample thickness no longer occurred below a certain sample thickness, a result which probably corresponds to a transition from internal-resistance rate control to external-resistance rate control in their experimental apparatus. The velocity of the sublimation front would be independent of sample thickness for freeze-drying rate-limited by external resistances to heat and/or mass transfer.

4. The important temperature variable influencing the volatiles retention during freeze-drying is the *temperature of the sublimation front.* The effect of changes in the sublimation front temperature should reflect a compromise between the low D obtainable from the higher solids contents of the concentrate microregions which should prevail at lower frozen zone temperatures, on the one hand, and the low t obtainable from faster velocities of the sublimation front at higher temperatures. Such a dual effect is in qualitative agreement with the results of Flink and Karel[243] and Sauvageot et al.[242] Flink and Karel[243] have noted that the sublimation front temperature is closely related to the question of whether or not the substance will collapse upon freeze-drying. As noted earlier, collapse will give greater volatiles loss.

5. A number of investigators[86,94,242] have found that *species with markedly different relative volatilities in solution have similar retentions* during freeze-drying. This result would follow from the fact that the diffusivities of different trace organic compounds in aqueous carbohydrate solutions of a given composition should be much the same as one another. As noted by Menting, Hoogstad, and Thijssen,[244] the volatiles loss should be dependent upon diffusivity and independent of the relative volatility of the compound as long as the internal mass transfer resistance to volatiles migration is an order of magnitude or more greater than the external mass transfer resistance to volatiles loss.

6. For freeze-drying rate-limited by external resistances to heat and/or mass transfer, the sublimation front should move at a constant rate through the specimen. Consequently, the *degree of volatiles loss should be the same for different macroscopic regions* of the specimen. These pre-dictions are borne out by the experimental results of Flink and Karel,[94] whose studies on this question were carried out under conditions of external resistance to heat transfer being rate-controlling. When the rate of freeze-drying becomes limited by internal resistances to heat and mass transfer some gradient of residual volatiles content throughout the specimen could be expected, since the rate of passage of the sublimation front will be different at different locations.

Menting, Hoogstad, and Thijssen[245] have reported measurements of diffusivities for water and various organic species in aqueous maltodextrin solutions, covering a wide range of maltodextrin contents. Diffusivities for both water and the organic species are interpreted from an equivalent binary diffusion analysis and show the same sort of behavior illustrated in Figures 11a and 11b. The same authors[244] have compared predictions based upon these data with observed losses of the same volatile organic species during air-drying of water-maltodextrin slabs. They have also given approximate criteria for conditions which should give appreciable volatiles retention and for conditions which should give high volatiles retention during evaporative drying of homogeneous material.[244]

Chandrasekaran and King[246,247] have treated the volatiles loss problem through a three-component diffusion analysis, where the components are (1) the volatile organic species under consideration, (2) water, and (3) the dissolved solids, usually carbohydrates. Such an analysis is necessitated by the very different fluxes of water and of dissolved solids during drying, and by the fact that the relative fluxes of water and dissolved solids under drying conditions are very different from the relative fluxes of water and dissolved solids under conditions used for the measurement of diffusion coefficients. Chandrasekaran and King[246,247] report multicomponent diffusion coefficients for solutions composed of (1) ethyl alcohol, ethyl acetate, n-butyl acetate, or n-hexanal, (2) water, and (3) fructose and/or sucrose. Activity coefficients of the organic compounds in solution were also measured as a function of sugar content; these activity coefficients increase sharply with increasing sugar content. The diffusion measurements cover a range of sugar contents and were obtained by means of diaphragm cell experiments with two independent composition monitors. Three of the multicomponent diffusion coefficients are independent; the fourth is obtained by means of the Onsager reciprocal relationships. Substitution of these measured multicomponent diffusivities into the appropriate multicomponent diffusion equations, followed by computer solution of those equations for typical evaporative drying situations, shows that the terms involving cross-diffusion coefficients and interactive effects between components are indeed important. One interesting effect comes from the impact of the sugar-water composition gradient upon the gradient in chemical potential for the volatile organic species, stemming from the variation of the activity coefficient of the volatile organic species with respect to sugar content. This effect opposes the natural concentration gradient of the volatile organic species during a drying situation and results in a negative term in the flux equation for the volatile organic. This negative term, in turn, can provide a barrier against transport of the volatile organic and thereby provides another mechanism through which the loss of volatile species during drying can be controlled.

Processing Approaches

Ginnette and Kaufman[233] have reviewed design considerations for conventional processing systems. Warman and Reichel[248] discuss the factors to be considered in the development of freeze-drying processes and plants; they also have outlined the succession of processing steps required and possible means for implementing them, including raw material selection, preprocessing, freezing, drying, filling, and packaging. Oetjen and Eilenberg[249, 250] have described a means of freeze-drying from vibrated, or mixed, particulate beds and give information on continuous freeze-drying equipment developed in Germany.

Tooby[251] has put forward a freeze-drying process using moderate pressures of a circulating inert gas with desiccant for collection of the evolved water vapor. This process is similar to the circulating gas process proposed by Kan and deWinter[33, 185] and requires a very large gas circulation rate since an element of gas passes through or across the entire mass of food material before encountering the desiccant. In the layered or mixed bed process described earlier in this review[25, 26, 64, 198] the gas circulation rate required is reduced by interspersing desiccant and food material such that an element of the circulat-

ing gas encounters them sequentially and alternately.

The status of microwave approaches to freeze-drying has been comprehensively reviewed by Parker[252] and by Decareau.[253]

Freeze-Drying of Foamed Material

Foaming of liquid foods before or during freezing prior to freeze-drying has been recommended by several different investigators,[254-258] and in some instances is used in practice. Foaming before freeze-drying is a means of controlling product density and also gives a higher internal mass transfer coefficient during freeze-drying. The higher mass transfer coefficient may make it possible to dry from a solution of higher initial dissolved solids content than would otherwise be possible and may increase the drying rate unless the insulating properties of the dried foam layer become an important heat transfer rate limit. Two techniques for foaming material to be frozen and freeze-dried are described by Warman and Reichel.[248]

Hetzendorf and Moshy[259] have reviewed techniques and problems associated with drying from the liquid foamed state, such as in vacuum-puff-drying and in foam-mat drying.

Additives to Improve Freeze-Drying

Liquid foods rich in fructose and, to a lesser extent, glucose are particularly difficult to freeze-dry because of the tendency of the material to collapse into a viscous or glassy state. Thus, apple juice, grape juice, and honey are highly resistant to freeze-drying by ordinary methods.

Some time ago, Notter, Brekke, and Taylor[260] found that the addition of substantial quantities of sucrose or glucose to apple juice promoted good vacuum-puff-drying, avoiding the tendency which natural apple juice has to collapse during vacuum-puff-drying. For freeze-drying Stern and Storrs[257] recommend adding substantial quantities of either lactose or corn syrup solids to fruit juices rich in fructose or invert sugar so as to prevent loss of structure. They suggest that the amount of added solids be at least 75% of the amount of fructose in the juice. Moy[258] found that several tropical fruit juices could be converted into a powder by vacuum-puffing, followed by freeze-drying, providing sucrose was added before drying to bring the solids content to 40%. The addition of certain calcium compounds was also useful. In operating the atmospheric pressure fluidized bed freeze-drying process described earlier, Malecki et al.[80, 220] found that the addition of large quantities of orange juice powder to frozen apple juice particles reduced tendencies for particles to cake together and prevent fluidization. All of these instances involve adding large amounts of some other substance to the material being dried, thus altering the characteristics of the ultimate freeze-dried product.

Compressed Foods

Freeze-dried foods usually have well-preserved flavor and structure, and they weigh much less than did the original food before drying. One drawback from the standpoint of storage and transportation is that freeze-dried foods are not reduced in volume. In an effort to accomplish a reduced volume of the freeze-dried product there has been interest in compressing the freeze-dried product so as to eliminate most or all of the void space. Work of the U.S. Army on compressed freeze-dried foods has been reported by Brockmann,[234] and by Rahman, Henning, and Westcott.[261] Freeze-dried materials compressed in volume by up to a factor of 16 have been found to rehydrate remarkably well. It has been found, however, that it is necessary for the material to have a uniform moisture content of 5 to 10% before compression, so as to give the solid adequate plasticity. This moisture content has been achieved by controlled humidification of the freeze-dried product. MacKenzie and Luyet[262] have investigated ways of carrying out the original freeze-drying so as to stop at the moisture content desired for subsequent compression.

Acknowledgment

T. P. Labuza provided a helpful review and constructive suggestions, for which the author is grateful.

ARTICLES REVIEWED

A. Karel, M. and Flink, J., Mechanism of Volatile Retention in Freeze-Dried Carbohydrate Systems, presented at A.I.Ch.E. Meeting, Washington, D.C., November, 1969, *Chem. Eng. Progr. Symp. Ser.,* in press.

B. Kluge, G. and Heiss, R., Untersuchungen zur besseren Beherrschung der Qualität von getrockneten Lebenmitteln unter besonderer Berucksichtigung der Gefriertrocknung, *Verfahrenstechnik,* 1, 251 (1967).

C. Luikov, A. V., Heat and mass transfer in freeze-drying at high vacuum, in Symposium on Thermodynamic Aspects of Freeze-Drying, International Institute of Refrigeration, Commission X, Lausanne, Switzerland, 1969.

D. Rey, L. R. and Bastien, M.—C., Biophysical aspects of freeze-drying, in Freeze-Drying of Foods, Fisher, F. R., Ed., National Academy of Sciences—National Research Council, Washington, D. C., 1962.

E. Sandall, O. C., King, C. J., and Wilke, C. R., "The Relationship Between Transport Properties and Rates of Freeze-Drying of Poultry Meat," *A.I.Ch.E. J.,* 13, 428 1967; *Chem. Eng. Progr. Symp. Ser.,* 64, No. 86, 43 1968.

F. Thijssen, H. A. C. and Rulkens, W. H., Effect of freezing rate on rate of sublimation and flavor retention in freeze-drying, in Symposium on Thermodynamic Aspects of Freeze-Drying, International Institute of Refrigeration, Commission X, Lausanne, Switzerland, 1969.

REFERENCES

1. Flosdorf, E. W., *Freeze-Drying,* Reinhold, New York, 1949.

2. Noyes, R., Freeze Drying of Foods and Biologicals, Food Processing Review No. 1, Noyes Development Corp., Park Ridge, N. J., 1968.

3. Cerre, P. and Mestre, E., La lyophilisation des effluents radio-actifs, in *Aspects Théoriques et Industriels de la Lyophilisation,* Rey, L., Ed., Hermann, Paris, 1964.

4. Rey, L., Orientations nouvelles de la lyophilisation, in *Aspects Théoriques et Industriels de la Lyophilisation,* Rey, L., Ed., Hermann, Paris, 1964.

5. Rowe, T. W. G., Energy, mass transfer and economy in large scale freeze-drying, in *Aspects Théoriques et Industriels de la Lyophilisation,* Rey, L., Ed., Hermann, Paris, 1964.

6. Rey, L., Fundamental aspects of lyophilization, in *Aspects Théoriques et Industriels de la Lyophilisation,* Rey, L., Ed., Hermann, Paris, 1964.

7. Van Arsdel, W. B., *Food Dehydration,* Vol. I, Avi Pub. Co., Westport, Conn., 1964.

8. Cotson, S. and Smith, D. B., *Freeze-Drying of Foodstuffs,* Columbine Press, Manchester, England, 1962.

9. Harper, J. C. and Tappel, A. L., in *Advances in Food Research,* Vol. 7, Stewart, G. F., and Mrak, E. M., Eds., Academic Press, New York, 1957.

10. Burke, R. F. and Decareau, R. V., in *Advances in Food Research,* Vol. 13, Chichester, C. O., Mrak, E. M., and Stewart, G. F., Eds., Academic Press, New York, 1964.

11. Corridon, G. A., Freeze-Drying of Foods. A List of Selected References, U.S. Department of Agriculture, National Agricultural Library, List No. 77, 1964.

12. Peck, R. E. and Wasan, D. T., Progress in drying fundamentals, in *Advances in Chemical Engineering,* Drew, T. B., Hoopes, J., and Vermeulen, T., Eds. in press.

13. Luikov, A. V., in Proceedings of the 2nd International Congress of Food Science and Technology, Tilgner, D. J., and Borys, A., Eds., Warsaw, 1966.

14. Fulford, G. D., *Can. J. Chem. Eng.,* 47, 378 1969.

15. Meffert, H. F. T., *Dechema-Monographien, 63,* 127, 1969.

16. Meffert, H. F. T., in Heat Transmission Freeze-Drying Materials, Proceedings XI International Congress on Refrigeration, Munich (1963). Includes Supplementary Appendix, Sprenger Instituut, Wageningen, The Netherlands, 1963.

17. Bralsford, R., *J. Food Technol.,* 2, 339, 353, 1967.

18. Brajnikov, A. M., Vassiliev, A. I., Voskoboinikov, V. A., and Kautchechvili, E. I., Transfer de chaleur et de masse dans les matériaux poreux pendant la lyophilisation sous vide, in Symposium on Thermodynamic Aspects of Freeze-Drying, International Institute of Refrigeration, Commission X, Lausanne, Switzerland, 1969.

19. Luikov, A. V., Heat and Mass Transfer in Drying Processes, Gosenergoizdat, Moscow, 1956.

20. Luikov, A. V. and Vasiliev, L. L., Heat transfer of capillary porous bodies in a rarefied gas flow, in Symposium on Thermodynamic Aspects of Freeze-Drying, International Institute of Refrigeration, Commission X, Lausanne, Switzerland, 1969.

21. Hardin, T. C., Ph.D. Thesis, Georgia Institute of Technology, Atlanta, 1965.

22. Margaritis, A. and King, C. J., paper presented at A.I.Ch.E. Meeting, Portland, Oregon, 1969; *Chem. Eng. Progr. Symp. Ser,* in press.

23. Beke, G., The effect of the sublimation temperature on the rate of the freeze-drying process and upon the volumetric change in meat muscle tissue, in Proceedings of the XII International Congress of Refrigeration, Vol. 3, Madrid, 1969.

24. Clark, J. P., Ph.D. Thesis, University of California, Berkeley, 1968.

25. King, C. J. and Clark, J. P., *Food Technol.,* 22, 1235, 1968.

26. Clark, J. P. and King, C. J., paper presented at A.I.Ch.E. Meeting, Los Angeles, Calif., 1968; *Chem. Eng. Progr. Symp. Ser.,* in press.

27. Hatcher, J. D., M. S. Thesis, Georgia Institute of Technology, Atlanta, Georgia, 1964.

28. MacKenzie, A. P., *Ann. N.Y. Acad. Sci.,* 125, 522 1965.

29. MacKenzie, A. P., *Bull. Parenteral Drug Ass.,* 20, 101, 1966.

30. Sandall, O. C., Ph.D. Thesis, University of California, Berkeley, 1966.

31. King, C. J., Lam, W. K., and Sandall, O. C., *Food Technol.,* 22, 1302, 1968.

32. Triebes, T. A. and King, C. J., *Ind. Eng. Chem. Process Des. Develop.,* 5, 430, 1966.

33. Kan, B. and deWinter, F., *Food Technol.,* 22, 1269, 1968.

34. Strasser, J., Heiss, R., and Görling, P., *Kältetechnik-Klimatisierung,* 18, 286, 1966; see also Strasser, J., Thesis, T. H., München, 1965.

35. Lambert, J. B. and Marshall, W. R. Jr., in Freeze-Drying of Foods, Fisher, F. R., Ed., National Academy of Sciences-National Research Council, Washington, D.C., 1962.

36. Novikov, P. A., *Int. Chem. Eng.,* 2, 174, 1962.

37. Smol'skii, B. M. and Novikov, P. A., *Int. Chem. Eng.,* 3, 203 1963.

38. Vagner, E. A. and Novikov, P. A., *Inzh. Fiz. Zhurn.,* 15, 788, 1968.

39. Gunn, R. D., Ph.D. Thesis, University of California, Berkeley, 1967.

40. Gunn, R. D. and King, C. J., paper presented at A.I.Ch.E. Meeting, Los Angeles, Calif., 1968; *Chem. Eng. Progr. Symp. Ser.,* in press.

41. Dyer, D. F. and Sunderland, J. E., *J. Heat Transfer,* 89, 109, 1967.

42. Dyer, D. F. and Sunderland, J. E., *J. Heat Transfer,* 90, 379, 1968.

43. Dyer, D. F., The influence of varying interface temperature on freeze-drying, in Symposium on Thermodynamic Aspects of Freeze-Drying, International Institute of Refrigeration, Commission X, Lausanne, Switzerland, 1969.

44. Luikov, A. V., *Heat and Mass Transfer in Capillary Porous Bodies,* Pergamon Press, Oxford, 1966.

45. Ginnette, L. F., Graham, R. P., and Morgan, A. I. Jr., *Natl. Symp. on Vacuum Technol., Trans.,* 5, 268, 1958.

46. Kessler, H. G., *Chem.-Ing. Tech.,* 34, 163, 1962.

47. Lusk, G., Karel, M., and Goldblith, S. A., *Food Technol.,* 18, 1625, 1964.

48. Massey, W. M. Jr. and Sunderland, J. E., *Food Technol.,* 21, 408, 1967.

49. Gaffney, J. J. and Stephenson, K. W., *Trans ASAE,* 11, 874, 1968.

50. Magnussen, O. A., Measurements of heat and mass transfer coefficients during freeze-drying, in Symposium on Thermodynamic Aspects of Freeze-Drying, International Institute of Refrigeration, Commission X, Lausanne, Switzerland, 1969.

51. Sharon, Z. and Berk, Z., Freeze-drying of tomato juice and concentrate: Studies on heat and mass transfer, in Symposium on Thermodynamic Aspects of Freeze-Drying, International Institute of Refrigeration, Commission X, Lausanne, Switzerland, 1969.

52. Harper, J. C., *A.I.Ch.E. J.,* 8, 298, 1962.

53. Harper, J. C. and El Sahrigi, A F., *Ind. Eng. Chem. Fundam.,* 3, 318, 1964.

54. Karel, M., Physical and chemical considerations in freeze-dehydrated foods, in *Exploration in Future Food-Processing Techniques,* Goldblith, S. A., Ed., M.I.T. Press, Cambridge, Mass., 1963, 54.

55. Lusk, G., Karel, M., and Goldblith, S. A., *Food Technol.,* 19, 620, 1965.

56. Carl, K. R. and Stephenson, K. W., *Trans. ASAE,* 8, 414 1965.

57. Haugh, C. G., Huber, C. S., Stadelman, W. J., and Peart, R. M., *Trans. ASAE,* 11, 877, 1968.

58. Hamre, M. L., Ph.D. Thesis, Purdue University, Lafayette, Indiana, 1966.

59. Saravacos, G. D. and Pilsworth, M. N., *J. Food Sci.,* 30, 773, 1965.

60. Spiess, W. E. L., Seiler, R. S., and Brinckmann, A., in Proceedings of the XII International Congress of Refrigeration, Vol. 3, Madrid, 1969.

61. Cowart, D. G., M.S. Thesis, Pennsylvania State University, University Park, Penna., 1964.

62. Mason, E. A., Malinauskas, A. P., and Evans, R. B., *J. Chem. Phys.,* 46, 3199, 1967.

63. Gunn, R. D. and King, C. J., *A.I.Ch.E. J.,* 15, 507, 1969.

64. Gunn, R. D., Clark, J. P., and King, C. J., Mass transport in freeze-drying: Basic studies and processing implications, in Symposium on Thermodynamic Aspects of Freeze-Drying, International Institute of Refrigeration, Commission X, Lausanne, Switzerland, 1969.

65. Spiess,, W. E. L., Wolf, W., Tirtohusodo, H., and Sole, C. P., The influence of the structure on the mass transfer in freeze-drying, in Symposium on Thermodynamic Aspects of Freeze-Drying, International Institute of Refrigeration, Commission X, Lausanne, Switzerland, 1969.

66. Kramers, H. Rate-controlling factors in freeze-drying, in *Fundamental Aspects of the Dehydration of Foodstuffs,* Soc. Chem. Ind., Macmillan, New York, 1958.

67. Quast, D. G. and Karel, M., *J. Food Sci.,* 33, 170, 1968.

68. Monzini, A. and Maltini, E., Studies on the freeze-drying of frozen concentrated orange juice, in Symposium on Thermodynamic Aspects of Freeze-Drying, International Institute of Refrigeration, Commission X, Lausanne, Switzerland, 1969.

69. King, C. J., *Food Technol.,* 22, 165, 1968.

70. Koonz, C. H. and Ramsbottom, J. M., *Food Research,* 4, 120, 1939.

71. Kuprianoff, J., Fundamental and Practical Aspects of the Freezing of Foodstuffs, in *Aspects Theoriques et Industriels de la Lyophilisation,* Rey, L., Ed., Hermann, Paris, 1964.

72. Luyet, B. J., in Freeze-Drying of Foods, Fisher, F. R., Ed., National Academy of Sciences—National Research Council, Washington, D.C., 1962.

73. Rapatz, G. and Luyet, B., *Biodynamica,* 8, 121, 1959.

74. Lam, W. K., M.S. Thesis, University of California, Berkeley, 1967.

75. Rey, L., *Ann. N.Y. Acad. Sci.,* 85, 510 1960.

76. Lusena, C. B., *Ann. N.Y. Acad. Sci.,* 85, 541 1960.

77. Luyet, B. J., *J. Phys. Chem.,* 43, 881, 1939.

78. Luyet, B. J., *The Vitreous State of Matter, Its Measurement and Control,* Reinhold, New York, 1941.

79. Luyet, B. J., *Ann. N.Y. Acad. Sci.,* 85, 549, 1960.

80. Malecki, G. J., Shinde, P., Morgan, A. I. Jr., and Farkas, D., *Food Technol.,* 24, 601, 1970.

81. Cruz Picallo, J. A., Freeze-drying of coffee extracts, in Proceedings of the XII International Congress of Refrigeration, Vol. 3, Madrid, 1969.

82. Elerath, B. E. and Pitchon, E. (General Foods Corp.), U.S. Patent 3,373,042, March 12, 1968.

83. Johnson, J. W., Ponzoni, G. B., and Clinton, W. P., U.S. Patent 3,244,529, April 5, 1966.

84. Issenberg, P., Greenstein, G., and Boskovic, M., *J. Food Sci.,* 33, 621, 1968.

85. Saravacos, G. D. and Moyer, J. C., *Chem. Eng. Progr. Symp. Ser.,* 64, No. 86, 37, 1968.

86. Chandrasekaran, S. K. and King, C. J., paper presented at A.I.Ch.E. Meeting, Washington, D.C., 1969, *Chem. Eng. Progr. Symp. Ser.,* in press.

87. Spiess, W. E. L., Dissertation, Universität Karlsruhe (T.H.), Karlsruhe, 1969.

88. Sole, P. and Spiess, W. E. L., *Verfahrenstechnik,* 3, 340 1969.

89. Thijssen, H. A. C. and Rulkens, W. H., *De Ingenieur,* CH45, November 22, 1968.

90. Thijssen, H. A. C. and Rulkens, W. H., paper presented at CHISA Conference, Marienbad, Czechoslovakia, 1969.

91. Menting, L. C., Ph.D. Thesis, Technological University Eindhoven, Eindhoven, Netherlands, 1969.

92. Rulkens, W. H. and Thijssen, H. A. C., *Trans. Inst. Chem. Eng.* (London), 47, T 292, 1969.

93. Chandrasekaran, S. K., unpublished data, Department of Chemical Engineering, University of California, Berkeley, 1969.

94. Flink, J. and Karel, M., *J. Agr. Food Chem.,* 18, 295, 1970.

95. Karel, M., personal communication, 1969.

96. Schultz, T. H., Dimick, K. P., and Makower, B., *Food Technol.,* 10, 57, 1956.

97. Spiess W. E. L., *Kältetechnik,* 16, 349, 1964.

98. Rockland, L. B., *Food Technol.,* 23, 1241, 1969.

99. Nemitz, G., *ASHRAE J.,* 7, No. 3, 68, 1965.

100. Rey, L., L'humidité résiduelle des produits lyophilisés. Nature—origine et méthodes d'étude, in *Aspects Théoriques et Industriels de la Lyophilisation,* Rey, L., Ed., Hermann, Paris, 1964.

101. Simatos, D., L'eau et les formes de liaison de l'eau dans les produits lyophilisés, in *Aspects Théoriques et Industriels de la Lyophilisation,* Rey, L., Ed., Hermann, Paris, 1964.

102. Nemitz, G., Dissertation, Technische Hochschule, Karlsruhe, 1961.

103. Riedel, L., *Kältetechnik,* 8, 374, 1956.

104. Riedel, L., *Kältetechnik,* 9, No. 2, 38, 1957.

105. Labuza, T. P., *Food Technol.,* 22, No. 3, 15, 1968.

106. Loncin, M., Bimbenet, J. J., and Lenges, J., *J. Food Technol.,* 3, 131, 1968.

107. Salwin, H., *Food Technol.,* 17, No. 9, 34, 1963.

108. Saravacos, G. D., *Food Technol.,* 19, No. 4, 193, 1965.

109. Saravacos, G. D. and Stinchfield, R. M., *J. Food Sci.,* 30, 779, 1965.

110. MacKenzie, A. P. and Luyet, B. J., Cryobiology, 3, 341, 1967.

111. Strasser, J., *J. Food Sci.,* 34, 18, 1969.

112. Goldblith, S. A. and Tannenbaum, S. R., The nutritional aspects of the freeze-drying of foods, in Proceedings of the 7th International Congress of Nutrition, Vol. 4, Friedr. Vieweg & Sohn, Braunschweig, 1966.

113. Goldblith, S. A., Karel, M., and Lusk, G., The role of food science and technology in the freeze dehydration of foods, in *Aspects Théoriques et Industriels de la Lyophilisation,* Rey, L., Ed., Hermann, Paris. 1964.

114. Goldblith, S. A., Freeze-dehydration of foods, in *Aspects Théoriques et Industriels de la Lyophilisation,* Rey, L., Ed., Hermann, Paris, 1964.

115. Bimbenet, J. J. and Guilbot, A., *Chim. Ind.—Genie chimique,* 96, No. 4, 1966.

116. Lundberg, W. O., Mechanisms of lipid oxidation, in Schultz, H. W., Day, E. A., and Sinnhuber, R. O., *Lipids and Their Oxidation,* Avi Pub. Co., Westport, Conn., 1962.

117. Martinez, F. and Labuza, T. P., *J. Food Sci.,* 33, 241, 1968.

118. Karel, M., Ph.D. Thesis, Massachusetts Institute of Technology, 1960.

119. Goldblith, S. A., Karel, M., and Lusk, G., *Food Technol.,* 17, 139, 258, 1963.

120. Bengtsson, O. and Bengtsson, N. E., *J. Sci. Food Agri.,* 19, 317, 481, 486 1968.

121. Maloney, J. F., Labuza, T. P., Wallace, D. H., and Karel, M., *J. Food Sci.,* 31, 878, 1966.

122. Labuza, T. P., Maloney, J. F., and Karel, M., *J. Food Sci.,* 31, 885, 1966.

123. Lea, C. H., Chemical changes in the preparation and storage of dehydrated foods, in *Fundamental Aspects of the Dehydration of Foodstuffs,* Soc. Chem. Ind., Macmillan, New York, 1958.

124. Hodge, J. E., *J. Agr. Food Chem.,* 1, 928, 1953.

125. Ellis, G. P., in *Advances in Carbohydrate Chemistry,* Vol. 14, Wolfrom, M. L. and Tipson, R. S., Eds., Academic Press, New York, 1959.

126. Reynolds, T. M., in *Advances in Food Research,* Vol. 14, Chichester, C. O., Mrak, E. M., and Stewart, G. F., Eds., Academic Press, New York, 1963.

127. Song, P. S., Chichester, C. O. and Stadtman, F. H., *J. Food Sci.,* 31, 906, 1966; Song, P. S. and Chichester, C. O., *J. Food Sci.,* 31, 914, 1966; *J. Food Sci.,* 32, 98, 107, 1967.

128. Acker, L., Enzymatic reactions in foods of low moisture content, in *Advances in Food Research,* Vol. 11, Chichester, C. O., Mrak, E. M., and Stewart, G. F., Eds., Academic Press, New York, 1961.

129. Acker, L. W., *Food Technol.,* 23, 1257, 1969.

130. Lund, B. D., Fennema, O. and Powrie, W. D., *J. Food Sci.,* 34, 378, 1969.

131. Lusk, G., Karel, M., and Goldblith, S. A., *Food Technol.,* 18, 157, 1964.

132. Bengtsson, N. E., *J. Food Technol.,* 2, 365, 1967.

133. MacKenzie, A. P. and Luyet, B. J., *Nature,* 215, No. 5096, 83, 1967.

134. Schultz, H. W. and Anglemeir, A. F., Eds., *Proteins and Their Reactions,* Avi Pub. Co., Westport, Conn., 1964.

135. Aitken, A., *J. Food Sci. Agr.,* 13, 439, 1962.

136. Miller, W. O. and May, K. N., *Food Technol.,* 19, 1171, 1965.

137. Tuomy, J. M., Schlup, H. T., and Helmer, R. L., *Food Technol.,* 23, 334, 1969.

138. Hamm, R., Biochemistry of Meat Rehydration, in *Advances in Food Research,* Vol. 10, Chichester, C. O., Mrak, E. M. and Stewart, G. F., Eds., Academic Press, New York, 1960.

139. Spicer, A., *Food Technol.,* 23, 1272, 1969.

140. Bengtsson, N. E., Rapport 230, Svenska Institutet för Konserveringsforskning, Göteborg, Sweden, 1967.

141. Bengtsson, N. E., Rapport 232, Svenska Institutet för Konserveringsforskning, Göteborg, Sweden, 1967.

142. Tease, S. C., Progress of food freeze-drying in the the U.S.A., in *Aspects Théoriques et Industriels de la Lyophilisation,* Rey L., Ed., Hermann, Paris, 1964.

143. Smith, D. B., Commercial equipment in the United States, in Cotson, S., and Smith, D. B., *Freeze-Drying of Foodstuffs,* Columbine Press, Manchester, England, 1962.

144. Bird, K., Freeze-Drying of Foods: Cost Projections, Marketing Research Report No. 639, U.S.D.A. Economic Research Service, 1964.

145. Bird, K., The Awakening Freeze-Drying Industry, USDA Economic Research Service, 1965.

146. Kroll, K., *Aufbereitungs-technik,* 4, 287, 1964.

147. Miner, S. M., *ASHRAE J.,* 7, No. 6, 92, 1965.

148. Abbott, J. A. and Thuse, E. (FMC Corp.), U.S. Patent 3,132,930, May 12, 1964.

149. Lorentzen, J. (Atlas A/S), U. S. Patent 3,382,586, May 14, 1968.

150. Marich, F., *Food Eng.,* 37, No. 4, 52, No. 6, 85, 1965.

151. Dalgleish, J. M., Leybold continuous plant equipment, in Cotson, S., and Smith, D. B., *Freeze-Drying of Foodstuffs,* Columbine Press, Manchester, England, 1962.

152. Hackenberg, U., U.S. Patent 3,273,259, September 20, 1966.

153. Rockwell, W., Kaufman, V., and Lowe, E. (USDA), U.S. Patent 3,303,578, February 14, 1967.

154. Rockwell, W., Kaufman, V., Lowe, E., and Morgan, A. I. Jr., *Food Eng.,* 37, No. 4, 49, 1965.

155. Kaufman, V. and Rockwell, W. (USDA), U.S. Patent 3,308,552, March 14, 1967.

156. Pfluger, R. A., Ewald, J. F., and Elerath, B. E. (General Foods Corp.), U.S. Patent 3,365,806, January 30, 1968.

157. Stinchfield, R. M. (Arthur D. Little, Inc.), U.S. Patent 3,218,731, November 23, 1965.

158. Fuentevilla, M. (Pennsalt Chemicals Corp.), U.S. Patent 3,264,747, August 9, 1966.

159. Rowe, T. W. G., Vacuum systems for freeze-drying, in Cotson, S., and Smith, D. B., *Freeze-Drying of Foodstuffs,* Columbine Press, Manchester, England, 1962, pp. 12 ff.

160. Rowe, T. W. G., Recent advances in vacuum methods, in *Aspects Théoriques et Industriels de la Lyophilisation,* Rey, L., Ed., Hermann, Paris, 1964.

161. Barrett, J. P., Laxon, R., and Webster, P. H. N., *Food Technol.,* 18, 38, 1964.

162. Blake, J. H., Pelmulder, J. P., and Thuse, E., (FMC Corp.), U.S. Patent 3,382,585, May 14, 1968.

163. Togashi, H. and Mercer, J. L., (Cryo-Maid, Inc.), U.S. Patent 3,247,600, April 26, 1966.

164. Tyson, R. (Pennsalt Chemicals Corp.), U.S. Patent 3,281,949, November 1, 1966.

165. Tucker, W. H. and Sherwood, T. K., *Ind. Eng. Chem.,* 40, 832, 1948.

166. Rowe, T. W. G., *Le Vide (Paris)*, 102, 516, 1962.

167. Thuse, E. (FMC Corp.), U.S. Patent 3,132,929, May 12, 1964.

168. Eolkin, D., (Gerber Products Co.), U.S. Patent 3,210,861, October 12, 1965.

169. Kumar, R., M.S. Thesis, University of California, Berkeley, 1968.

170. Saravacos, G. D., *Food Technol.*, 21, 187, 1967.

171. Strasser, J., presented at Annual Meeting, Institute of Food Technology, Minneapolis, Minn., 1967.

172. Graham, R. P., Brown, A. H., and Ramage, W. D., U.S. Patent 2,853,797, 1958.

173. Hanson, S. W. F., The Accelerated Freeze Drying (AFD) Method of Food Preservation, Ministry of Agriculture, Fisheries and Food, London, 1961.

174. Oldencamp, H. A. and Small, R. F. (FMC Corp.), U.S. Patent 3,199,217, August 10, 1965.

175. Smithies, W. R. and Blakley, T. S., *Food Technol.*, 13, 610, 1959.

176. Brynko, C. and Smithies, W. R. (Canadian Government), U.S. Patent 2,930,139, March 29, 1960.

177. Meryman, H. T., *Science*, 130, 628, 1959.

178. Meryman, H. T. (U.S. Navy), U.S. Patent 3,096,163, July 2, 1963.

179. Lewin, L. M. and Mateles, R. I., *Food Technol.*, 60, 94, 1962.

180. Woodward, H. T., *Food Eng.*, 35, No. 6, 96, 1963.

181. Sinnamon, H. I., Komanowsky, M., and Heiland, W. K., *Food Technol.*, 22, 219, 1968.

182. Ehlers, H., Hackenberg, U., and Oetjen, G. W., *Trans. Nat. Vacuum Symp.*, 8, 1069, 1961.

183. Oetjen, G. W., Ehlers, H., Hackenberg, U., Moll, J., and Neumann, K. H., Temperature measurements and control of freeze-drying processes, in Freeze-Drying of Foods, Fisher, F. R., Ed., National Academy of Sciences—National Research Council, Washington, D. C., 1962.

184. Harper, J. C., U.S. Patent 3,271,873, September 13, 1966.

185. Kan, B. (United Fruit Co.), U.S. Patent 3,263,335, August 2, 1966.

186. Larson, R. W., Steinberg, M. P., and Nelson, A. I., *Food Technol.*, 21, 401, 1967.

187. Thuse, E. (FMC Corp.), U.S. Patent 3,299,525, January 24, 1967.

188. Blake, J. H., Technical Report FD-30, Contract DA-19-129-AMC-369(N), U.S. Army Materiel Command, U.S. Army Natick Laboratories, Natick, Mass., 1965.

189. Barth, J. R., Pelmulder, J. P., and Thuse, E., (FMC Corp.), U.S. Patent 3,218,728, November 23, 1965.

190. Blake, J. H., Pelmulder, J., and Thuse, E. (FMC Corp.), U.S. Patent 3,382,584, May 14, 1968.

191. Mink, W. H. and Nack, H., U.S. Patent 3,239,942, March 15, 1966.

192. Dryden, C. E. and Nack, H., U.S. Patent 3,269,025, August 30, 1966.

193. Sachsel, G. F. and Mink, W. H., U.S. Patent 3,319,344, May 16, 1967.

194. Mink, W. H. and Sachsel, G. F., *Chem. Eng. Progr. Symp. Ser.*, 64, No. 86, 54, 1968.

195. deBuhr, J. (United Fruit Co.), U.S. Patent 3,262,212, July 26, 1966.

196. Kan, B. (United Fruit Co.), U.S. Patent 3,255,534, June 14, 1966.

197. Mellor, J. D., (Commonwealth Scientific and Industrial Research Org.), U.S. Patent 3,352,024, November 14, 1967.

198. King, C. J. and Clark, J. P. (USDA), U.S. Patent 3,453,741, July 8, 1969.

199. Wistreich, H. E. and Blake, J. A., *Science*, 138, 138, 1962.

200. Bohrer, B. (Sun Oil Co.), U.S. Patent 3,298,109, January 17, 1967.

201. Rey, L. (Air Liquide), U.S. Patent 3,271,875, September 13, 1966.

202. Copson, D. A., *Microwave Heating in Freeze-Drying, Electronic Ovens and Other Applications*, Avi Pub. Co., Westport, Conn., 1962.

203. Harper, J. C., Chichester, C. O., and Roberts, T. E., *Agr. Eng.*, 43, 78, 90, 1962.

204. Decareau, R. V., Microwave freeze-drying, in Cotson, S., and Smith, D. B., *Freeze-Drying of Foodstuffs*, Columbine Press, Manchester, England, 1962.

205. Horejsi, V. and Fric, V., Proceedings of the XI International Congress of Refrigeration, Munich, 1963, 1519.

206. Meryman, H. T., Induction and dielectric heating for freeze-drying, in *Aspects Théoriques et Industriels de la Lyophilisation*, Rey, L., Ed., Hermann, Paris, 1964.

207. Gouigo, E. I., Malkov, L. S., and Kaoukhtchechvili, E. I., Certaines particularités du transfert de chaleur et de masse au course de la lyophilisation de produits poreaux dans le champ ultra haute frequence, in Symposium on Thermodynamic Aspects of Freeze-Drying, International Institute of Refrigeration, Commission X, Lausanne, Switzerland, 1969.

208. Hoover, M. W., Markantonatos, A., and Parker, W. N., *Food Technol.*, 20, 807, 1966.

209. Hoover, M. W., Markantonatos, A., and Parker, W. N., *Food Technol.*, 20, 811, 1966.

210. Hammond, L. H., *Food Technol.*, 21, 735, 1967.

211. Ginzburg, A. S. and Lyakhovitsky, B. M., Using infrared radiation to intensify the freeze-drying process, in Proceedings of the XII International Congress of Refrigeration, Vol. 3, Madrid, 1969, 993.

212. Zamzow, W. H. and Marshall, W. R. Jr., *Chem. Eng. Progr.*, 48, 21, 1952.

213. Greaves, R. I. N., *J. Pharm. Pharmacol.*, 14, 621, 1962.

214. Hackenberg, U., U.S. Patent 3,234,658, February 15, 1966.

215. Seffinga, G. (Seffinga Engineering Co., NV.) U.S. Patent 3,264,745, August 9, 1966.

216. Nair, J. H., *Food Eng.*, 38, No. 6, 105, 1966.

217. Clarke, R. J., *Food Manufacture*, 41, No. 11, 42, 1966.

218. Greaves, R. I. N., High vacuum spray freeze-drying, in *Aspects Théoriques et Industriels de la Lyophilisation*, Rey, L., Ed., Hermann, Paris, 1964.

219. Thuse, E., Ginnette, L. F., and Derby, R. (FMC Corp.), U.S. Patent 3,362,835, January 9, 1968.

220. Malecki, G. (USDA), U.S. Patent 3,313,032, April 11, 1967.

221. Hutton, R. (U.S. Army), U.S. Patent 3,309,777, March 21, 1967.

222. Muller, J. G., *Food Technol.*, 21, 49, 1967.

223. Merson, R. L. and Morgan, A. I. Jr., *Food Technol.*, 22, 631, 1968.

224. Davis, J. M. and Pfluger, R. A. (General Foods Corp.), U.S. Patent 3,188,750, June 16, 1965.

225. Havelka, J., Etude du transfert de chaleur et de masse pendant la lyophilisation. Conditions pour une régulation automatique du procede plus particulierement dans les materiaux liquides, in Symposium on Thermodynamic Aspects of Freeze-Drying, International Institute of Refrigeration, Commission X, Lausanne, Switzerland, 1969.

226. Bouldoires, J. P., Etude experimentale des transferts de chaleur et de masse en course de lyophilisation par mesures dielectriques et par mesures de pressions de vapeur, in Symposium on Thermodynamic Aspects of Freeze-Drying, International Institute of Refrigeration, Commission X, Lausanne, Switzerland, 1969.

227. Rey, L. R., *Biodynamica*, 8, 241, 1961.

228. Rey, L. R., (Centre National de la Recherche Scientifique), U.S. Patent 3,078,586, February 26, 1963.

229. Neumann, K., Les problemes de mesure et de reglage en lyophilisation, in *Traite de Lyophilisation*, Rey, L., Ed., Hermann, Paris, 1961.

230. Neumann, K., U.S. Patent 2,994,132, August 1, 1961.

231. Neumann, K., (Leybold Hochvakuum Anlagen GMBH), U.S. Patent 3,077,036, February 12, 1963.

232. Illich, G. M. (Abbott Laboratories), U.S. Patent 3,259,991, July 12, 1966.

233. Ginnette, L. F. and Kaufman, V. F., Freeze-drying of foods, in *The Freezing Preservation of Foods,* Tressler, D. K., Van Arsdel, W. B., and Copley, M. J., Eds., 4th ed., Vol. 3, Avi Pub. Co., Westport, Conn., 1968, 377.

234. Brockmann, M. C., *Chem. Eng. Progr. Symp. Ser.*, 66, No. 100, 53, 1970.

235. Rothmayr, W., *Ind. Chim. Belge*, 34, 1089, 1969.

236. King, C. J., *Recent Developments in Food Dehydration Technology,* presented at 3rd International Congress of Food Science and Technology, Washington, D. C., 1970.

237. Schnettler, F. J., Monforte, F. R., and Rhodes, W. W., *Science of Ceramics*, 4, 1968.

238. Jason, A. C. and Jowitt, R., *Dechema-Monographien*, 63, 21, 1969.

239. Riedel, L., *Dechema-Monographien*, 63, 115, 1969.

240. Partmann, W., *Dechema-Monographien*, 63, 93, 1969.

241. Huber, C. S., Harrington, R. B., and Stadelman, W. J., *J. Food Sci.*, 35, 239, 233, 1970.

242. Sauvageot, F., Beley, P., Marchand, A., and Simatos, D., paper presented at Symposium on Surface Reactions in Freeze-Dried Systems, International Institute of Refrigeration, Commission X, and Soc. chim. ind., Paris, 1969.

243. Flink, J. and Karel, M., *J. Food Sci.*, 35, 444, 1970.

244. Menting, L. C., Hoogstad, B., and Thijssen, H. A. C., *J. Food Technol.*, 5, 127, 1970.

245. Menting, L. C., Hoogstad, B., and Thijssen, H. A. C., *J. Food Technol.*, 5, 111, 1970.

246. Chandrasekaran, S. K., Ph.D. Dissertation, Univ. of California, Berkeley, 1971.

247. King, C. J. and Chandrasekaran, S. K., paper presented at XIII International Congress of Refrigeration, Commission X, Washington, D. C., 1971.

248. Warman, K. G. and Reichel, A. J., *The Chemical Engineer*, CE134, May 1970.

249. Oetjen, G. W., Freeze-Drying Processes and Equipment developed or used in the Federal Republic of Germany, presented at XVIII Congresso Nazionale del Freddo, Padua, 1969.

250. Oetjen, G. W. and Eilenberg, H. J., Heat transfer during freeze-drying with moved particles, in Symposium on Thermodynamic Aspects of Freeze-Drying, International Institute of Refrigeration, Commission X, Lausanne, Switzerland, 1969.

251. Tooby, G., U. S. Patent 3,466,756, September 16, 1969.

252. Parker, W. N., in *Microwave Power Engineering*, Academic Press, New York, 1968.

253. Decareau, R. V., *CRC Critical Reviews in Food Technology*, 1, 199, 1970.

254. Ginnette, L. F., Lampi, R. A., and Abbott, J. A. (FMC Corp.), U. S. Patent 3,309,779, March 21, 1967.

255. Wertheim, J. and A. R. Mishkin (Nestle's Products, Ltd.), British Patent 1,102,587, February 7, 1968.

256. Pfluger, R. A., Schulman, M., and Hetzendorf, M. S. (General Foods Corp.), U. S. Patent 3,482,990, December 9, 1969.

257. Stern, R. M. and Storrs, A. B. (Great Lakes Biochemical Co., Inc.), U. S. Patent 3,483,032, December 9, 1969.

258. Moy, J. H., Vacuum Puff Freeze-Drying of Tropical Fruit Juices, presented at Institute of Food Technology Meeting, San Francisco, 1970.

259. Hetzendorf, M. S. and Moshy, R. J., *CRC Critical Reviews in Food Technology*, 1, 25, 1970.

260. Notter, G. K., Brekke, J. E., and Tyalor, D. H., *Food Technol.*, 13, 341, 1959.

261. Rahman, A. R., Henning, W. L., and D. E. Westcott, paper presented at Institute of Food Technologists meeting, San Francisco, 1970.

262. MacKenzie, A. P. and Luyet, B. J., Recovery of Compressed Dehydrated Foods, Tech. Rept. 70-16-FL, U. S. Army Natick Laboratories, Natick, Mass., 1969.